To Fred.   with best wishes
                            Elaine Morgan
More Best Wishes.
         from:
Priscilla M. Perkins.

            Dr Holley. - July 05
              Irene Perkins.

Love & thanks from

   Bev x 2005
     Oaktree Barn

PINKER'S LIST

PINKER'S LIST

Elaine Morgan

Copyright © 2005 Elaine Morgan
All rights reserved.

EILDON PRESS

ILLUSTRATIONS BY PETER BATCHELOR

ISBN 0 9525620 2 2

Printed in the United States of America

DEDICATION

"But principally," wrote Jonathan Swift, "I hate and detest that animal called man."

Disgust with the human race has seldom been voiced more passionately than it is today, and some even say the planet would be a better place if we had never appeared on it. We are all Swiftians now.

But the satirist did not stop there. Here is the rest of his sentence: ". . .although I heartily love John, Peter, Thomas, and so forth."

This book is therefore dedicated to John, Peter, Thomas and so forth, together with Mary, Catherine, Jane, and so forth, and their counterparts around the world.

ACKNOWLEDGEMENTS

I am indebted to Judith Hayes, who from half a world away in California sustained my faith in this project and rendered invaluable secretarial assistance.

Contents

1. The Invitation 9
2. Darwin and God 13
3. Darwin and Marx 21
4. Lamarckists... 29
5. The Twig Is Bent 39
6. Thicker than Water 49
7. Bread upon the Waters 59
8. The Troubles 75
9. Genes and Memes 85
10. The Pleistocene Inheritance 99
11. Cinderella 111
12. Rape 123
13. The Origin of Empathy...............................137
14. It's a Boy...151
15. Right and Left...173
16. Striding the Blast..191
17. Progress...207
18. What's Left?..223
19. The Magic Hand...237
20. The End of History?...................................255

References..273

Index..285

Trust me, I'm a robot.

Chapter 1
The Invitation

The greatest enterprise of the mind has always been and will always be the attempted linkage of the sciences and the humanities.

— Professor E. O. Wilson

The poet W. H. Auden once described his sensations on finding himself in the company of scientists. He felt, he said, "like a shabby curate who has strayed by mistake into a drawing room full of dukes." When Richard Dawkins quoted this, he added that he himself feels much the same way about poets.

It is an admirable human trait to express particular regard for those talents that we don't share. But if the sciences and the humanities keep tiptoeing around one another with quite so much reverence, there is little hope of fulfilling the dreams of another scientist, Professor E. O. Wilson.

In 1998 Wilson launched a campaign in favour of what he called consilience. He wanted to break down the barriers of over-specialisation and encourage people of all academic disciplines to exchange views. He invited artists, historians, poets, musicians, sociologists, theologians and philosophers to stroll with him in what used to be the No Man's Land between the sciences and the humanities. I found the idea instantly appealing, since I had attempted a cross-over from the arts to the sciences in mid-career.

Wilson's idea was not totally new. In England about forty years earlier, the physicist/novelist C. P. Snow had appealed for more fraternisation between what he called the Two Cultures. The difference was that in 1959 the scientists were on the defensive, newly arrived in the ancient seats of learning and prickly about being patronised. That pecking order has been reversed, so that it is now the older disciplines that are on the defensive. That may be one reason why there has been no rush to accept Wilson's invitation. Some have expressed fears that the rapprochement might turn into a take-over bid.

Their nervousness is understandable. In recent times scientists have acquired some amazing new technical tools which make it possible for them to compare the human brain with the computers that have been constructed in its image. Occasionally this leads to the assertion that the brain to all intents and purposes *is* a computer, and a human being is in the last resort a robot. As a stimulus to the imagination these are great gambits, but they create their own difficulties. If you tell me that all humans are robots, I must deduce that either you are mistaken or you are a robot, and that detracts a little from your credibility. "Trust me, I am a scientist" is a very potent assurance. "Trust me, I am a robot" doesn't have anything like the same ring about it.

Wilson did not specifically include politicians among the groups of people he hoped to welcome into his circle of consilience. Perhaps he should have done. If there is one area in which the boundaries between disciplines are most frequently trampled over, it is the interface between evolutionary science and politics. Professor Steven Pinker has recently chosen to visit this terrain, in his book *The Blank Slate*. He considers that his own academic training and clarity of thought have enabled him to throw a flood of light on political affiliations. He implies that one particular attitude to social and economic theories—one which he happens to share—has been scientifically validated, and should now be accepted by all fair-minded readers as the truth.

I accept his sincerity and respect his scholarship. But I believe there are other ways of looking at the same facts, which may lead to different conclusions. To illustrate this, I have tried in this book to present one of those different ways, and I began as Pinker did by retracing some of the propositions advanced by earlier thinkers.

The reader will not find this version any more objective than *The Blank Slate*, but I believe it is not less objective either. Science's most precious tool is scepticism. Since it is very hard for human beings to be totally sceptical of their own ideas, we must cherish our ability to be sceptical of one another's.

> **Summary** *When beauty is said to be in the eye of the beholder, few people want to argue. But scientists aspire to establish truths which are 100% uninfluenced by the eye of the beholder. In the study of human nature this aspiration faces its toughest challenge.*

"By next Friday morning, they will all be convinced they are monkeys."

Chapter 2
Darwin and God

Man has never been the same since God died.
He has taken it very hard. Why, you'd think it was only yesterday,
The way he takes it.

— E. St. V. Millay

In a sense it *was* only yesterday. It has happened substantially in the last two hundred years. And that is the blink of an eye, when we remember that formerly, as far back as thought can reach, the Immortals in one form or another dwelt in the collective imagination of every human community we have any knowledge of.

That does suggest that belief in the supernatural fulfilled some basic human need—and it is hardly surprising. From the day we are born throughout our most formative years, when our brains are being sculpted to fit optimally into our environment, we are surrounded by beings far bigger, stronger, and more knowledgeable than we are, with fixed ideas on how they want us to behave. One or two of them are likely to take a personal interest in us—a benevolent one, if we are lucky. Our modes of thinking and behaving are up to that point designed for living in a world in which that is the case.

But as we grow up the authority figures appear to dwindle until they are no bigger than ourselves, and no less fallible. Concern for our welfare no longer dominates so many of their working hours. In the end they are likely to disappear. We share this predicament with many other living things, especially other mammals, but it is more acute for humans since we have the longest period of dependency and there are more things we need to learn.

It may be less traumatic, then, if we can believe that our parents and more distant ancestors are not *really* dead, that somewhere or other they are still around keeping a benevolent eye on us. It might be better still if the role they played in our lives could be taken over by beings who are still more powerful and will never change or cease to exist. For many people this is the case: there is still life after death, and God is still in his heaven. But in much of the Western world the near-unanimity of belief in him has gone, and for many of the believers his previous personal relationship with them is undergoing a difficult reappraisal. The process is likely to spread to other parts of the world as time goes by, and that is bound to constitute a difficult rite of passage for the human race.

For a long time, one of the strongest intellectual grounds for continuing to believe in God was the wondrous diversity of the living world. How could it exist without a miraculously fertile creative imagination to bring it into being? "Poems are made by fools like me, but only God can make a tree." That is one reason why Darwin's theory of evolution caused such a shake-up in 1859. It did not disprove the existence of God—many believers have by now assimilated it as completely as they assimilated Copernican astronomy—but it has destroyed one of the strongest bulwarks against disbelief, simply by pointing out that "Evolution can make a tree." So Darwin's theory was initially greeted in England by most people of his own class with incredulity and horror, and he was not in the least surprised by their reaction. He had seen it coming only too clearly, since he had been a believer himself, and although he lost his faith he did not seem to experience the loss as a liberation. He was living in a Christian society and had no wish to hurt or offend his friends or his devout Unitarian wife. Even the scientists who corresponded with him were for the most part Christians, brought up to denounce the "transmutation of species" as a vile heresy.

Some of the attacks on him were the more embittered because by birth, education, and social connections, he had always been considered "one of us." After publishing *The Origin of Species* he was seen for a time as a renegade, offering aid and comfort to the enemies of the Established Church and all it stood for. Much of the opposition was pragmatic. Whereas Darwin was concentrating his mind on whether or not his hypothesis was sound, the protesters were more actively concerned about the probable *effect on society* of this undesirable idea, if it was allowed to get loose in the world. One of the most oft-quoted responses to the *Origin of Species* was fictional. "If this is true, let us hope that at least it will not become generally known" was the caption of a Punch cartoon depicting a society lady shocked by Darwin's book. But the words accurately reflected the feelings of a great many people at the time. It was less than fifty years since the defeat of Napoleon. To those of a nervous disposition it seemed quite possible that if the godless doctrines that had triggered the French Revolution were revived in this country, the streets of London might run red with blood.

True, there were a few clergymen, like the popular novelist Charles Kingsley, who never had the slightest difficulty in accepting evolution as part of the Divine plan. But there were others who resisted with the strength of desperation, like Philip

Gosse. He warned his countrymen that God was putting them on trial. It was obvious to Gosse that when He created the earth, He must have deliberately buried what looked like the fossilised bones of monsters beneath its surface, to test the faith of the palaeontologists who would one day dig them up.

The fears of devout and respectable people were not allayed by the way the evolution message was received at the other end of the social spectrum. Any hope that Darwin's theory might "not become generally known" had quickly evaporated, and among its most tireless promoters was T. H. Huxley. As soon as he read Darwin's book, he kicked himself for not having thought of such a simple and obvious idea for himself. He travelled tirelessly the length and breadth of Britain carrying the news to everyone who would listen, from academic seminars to evening classes in Mechanics' Institutes. He was himself surprised by the warmth of his reception in the latter. ("My working men stick by me wonderfully. By next Friday morning, they will all be convinced they are monkeys.")

England had been assumed to be a solidly Christian country, and few people had realised the extent of political atheism among the industrial working classes in the cities. Urban myths were being circulated about business men exhorted by the parson to dismiss any workers who were found to be free thinkers—only to find out that they comprised around ninety per cent of the workforce. For years they had been taught from the pulpit that they were miserable offenders, occupying their lowly positions by divine decree, full of original sin and liable to be punished for it after death by an eternity of torture if their faith faltered. Now suddenly Huxley was telling them that they were not the result of the Fall but of a Rise, a spectacular and incredible rise from the most unpromising beginnings. "Man," he wrote, "has worked his way to the headship of the sentient world and has become the superb animal he is in virtue of his success in the struggle for existence." And if he had risen so far already, to what heights might he not aspire in the years to come?

It was heady stuff. Huxley carried a large slice of the British public along with him because as well as being good at oratory, he was good at tactics. He was careful not to proclaim the death of God. He refused to take up any position as to whether God did or did not exist; in fact he invented a new word to describe his stance: "agnostic." One of his tours ended in a public meeting at the Albert Hall in London, which was full to capacity with an audience drawn from all sectors of society, and Huxley's inspirational eloquence was cheered to the echo.

How much the vilification of Darwin had been due to fear became clear at the time of his death twenty-three years later. The panic had proved groundless. Victoria still sat safely on her throne. Charles Darwin was forgiven, and buried in Westminster Abbey with all the pomp and ceremony due to a great Englishman. Today, the idea that species evolve by descent-with-modification is no longer controversial. There are still some Creationists, particularly active in America, who dispute it, but the accumulated evidence in its favour is generally regarded as overwhelming. Even Pope Pius XII in an encyclical letter in 1950 described 'evolutionism' as "a serious hypothesis, worthy of investigation."

As for its political impact, Darwin's ideas in his lifetime were widely seen as, on balance, progressive—resisted by conservatives, acclaimed by those who were critical of the society in which they found themselves and had visions of a better world.

They felt that the moral to be drawn from the evolutionary story was that Man had risen from humble beginnings—which augured well for those whose beginnings were still humble—and that all men were brothers.

> **Summary** *Darwin did not set out to disprove the existence of God, but* The Origin of Species *made it easier to imagine a world without a creator. It was originally welcomed by progressive thinkers largely because the Church's teachings were widely used to justify the status quo.*

...the book was found in his library with most of the pages uncut.

Chapter 3

Darwin and Marx

Darwin recognises among beasts and plants his English society.

— Karl Marx

One of the readers who welcomed the book was Karl Marx. In 1849, after publishing the *Communist Manifesto* and being expelled from Prussia, he had moved to London, and for the next three decades Darwin and Marx lived and worked barely 20 miles apart. Darwin, at Down House, was lovingly tended by Emma and protected against all unnecessary distraction. Marx lived in great poverty in two rooms in Soho with his wife and children and spent much of his time in the British Museum.

They had one or two things in common. Each was possessed by a central idea which dominated his life, and neither was deterred by the outrage with which their ideas were greeted. Very often they were attacked by the same people, and there is an old adage to the effect that "my enemy's enemy is my friend." But Darwin and Marx never met, and if they had, there would have been little chance of a rapprochement. Marx was too militant and Darwin was too bourgeois, and their intellectual interests were too far apart. Darwin was studying biology; Marx was studying social history and economics.

There was also a difference in their motivation. Marx was amassing facts which he hoped to use as a means of influencing people's behaviour and their future destiny. "The philosophers," he wrote, "have only interpreted the world in various ways; the point is to change it." Darwin was seeking to understand things, not to change them. It was not because his heart was not in the right place. Whenever he witnessed human suffering and injustice as he did in part of slave-owning America, he was deeply distressed. But he did not regard himself as being born to set it right.

Marx read *The Origin of Species* on its publication, and wrote to Engels about it: "Darwin's book is very important and it suits me well that it supports the class struggle in history from the point of view of natural science. One has, of course, to put up with the crude English method of discourse. Despite all deficiencies it not only deals the death blow to 'teleology' in the natural sciences for the first time but also sets forth the rational meaning in an empirical way."

On more mature consideration, he had some misgivings about the possible long-term effects of Darwin's reasoning. Marx's primary intention in life was to combat the idea of economists like Thomas Malthus, who preached the virtues of *laissez-faire*: i.e. unbridled capitalist competition. Malthus had published a warning that since human populations can expand much faster than the production of food can expand, it would be mistaken kindness to try to alleviate the sufferings of the poor by welfare provisions. They would simply reproduce themselves even faster and die in even greater misery.

It was obvious to Marx that Malthus's fingerprints were all over *The Origin of Species,* and he was right about that. Darwin himself recalled how the idea of natural selection had occurred to him shortly after reading Malthus's treatise on population—and A. R. Wallace, who independently arrived at the same idea while suffering from malaria in the middle of a monsoon in the East Indies, also connected his inspiration with memories of the same book.

Marx, though, remained convinced that Darwin's ideas, however arrived at, would on balance do more good than harm to his hopes of reforming society. Eight years after the publication of *Das Kapital*, he sent Darwin a copy of the book and received a courteous reply:

"Dear Sir,

> I thank you for the honour which you have done me by sending me your great work on Capital; and I heartily wish that I was more worthy to receive it by understanding more of the deep and important subject of political Economy. Though our studies have been so different, I believe that we both earnestly desire the extension of Knowledge, and that this is in the long run sure to add to the happiness of humankind.

I remain, dear Sir,
Yours faithfully,
Charles Darwin."

After Darwin's death, his copy of the book was found in his library with most of the pages uncut. His interest in politics was minimal, he did not find German easy to read, and while Marx could regard Darwin's ideas as useful to his own cause, the reverse was not true. Yet perhaps he read enough to gather the gist of it, because he was moved at one point to reflect: "If the misery of the poor be caused not by the laws of nature, but by our institutions, great is our sin." Before Marx, that thought would have been unlikely to cross the mind of anybody in Darwin's circle.

Neither man could have foreseen how many millions of words would be poured out in the 20th century in an attempt either to amalgamate Darwinism with Marxism or alternatively to vindicate one at the expense of the other. Both lived long enough to see their views distorted by people who claimed to be their supporters. When they were both in their graves, Darwin's name was used to defend Spencerian racism, and Marx's to try to justify the excesses of Stalinism.

For over forty years, published accounts of Marx's life recorded that he had made one further overture to Darwin—a proposition to dedicate *Das Kapital*, Vol 2, to him. This belief was based on a letter from Darwin found among papers in the possession of Marx's daughter Eleanor, stating that he would "prefer the Part or Volume not to be dedicated to me (though I thank you for the intended honour) as this implies to a certain extent my approval of the general publication, about which I know nothing." He added that "it has always been my object to avoid writing on religion, and I have confined myself to science." It was an odd letter, since volume two Capital is not about religion—and how could Darwin profess ignorance of the "general publication" when he was already in possession of the first volume?

In 1974 a neat piece of literary detective work revealed that the letter was in fact addressed to Edward Aveling, the man who after Marx's death became the lover of his daughter Eleanor. Aveling had asked to dedicate to Darwin a collection of essays entitled "The Students' Darwin"—one of the earliest popularisations of the work to introduce a racist thread which was never present either in Darwin's book or in Huxley's lectures on it.

Aveling was persistent in trying to recruit Darwin at least to the cause of militant atheism if not of communism. At a later date, he sent a telegram to Darwin to say that the renowned Dr Ludwig Buchner, President of the Congress of the International Federation of Free Thinkers, craved the honour of an interview with Darwin during his brief stay in London, and Aveling hoped to accompany him.

Darwin did not welcome visitors—he had previously explained "I am old and have very little strength,"—but he was incapable of being rude. The two atheists were invited to lunch. A Tory vicar, a friend of the family, was also invited, to sit between Emma and Aveling and ensure that she would not be outnumbered at her own table, where Charles discoursed at length on the neutral subject of worms. Later, in his study, he was induced to discuss his own loss of faith in God but no one could persuade him to proselytise it.

After Darwin's death it was often the Right rather than the Left which most frequently quoted his works to support their own political views. His cousin Francis Galton, who believed that mental traits were inherited to the same extent as physical ones, engaged in a campaign to increase the fitness of the human race by a process of "selective parenthood," and endowed a Chair of Eugenics at University College London. In the earliest days, eugenics was not seen as a politically loaded concept. There was widespread agreement that if some of the causes of human misery were hereditary, then any way of eliminating them deserved to be considered.

But later the idea was taken up—and in the event taken over—by people who had their own ideas about what constituted a desirable human being. Anxiety was often expressed about the fact that "the poor," as a class, were breeding faster than their betters. The economist Malthus had warned against aiding the unfit to survive and breed, and this theme was expanded into propositions for actively *preventing* them from doing so. Defective individuals, it was suggested, could be sterilised, humanely if possible but with no nonsense about giving them a choice in the matter. The term "defectives" suggested the insane and feeble-minded, but it proved to be elastic. Epileptics were included in the definition. Some categories of criminals could be added to the list. . .and how about "perverts" and women who kept having children out of wedlock?

The idea of editing *Homo sapiens* in this way seemed to become addictive. In England the concept of Social Darwinism was supported by Herbert Spencer and Walter Bagehot, and in America by W. G. Sumner and the American Eugenics Society. They felt that eugenics could do for society what natural selection had done for living organisms; the highest and the best could be selected and the others prevented from passing on their defects to future generations.

So sterilisation of the unfit was introduced—though not, oddly enough in England where much of the thinking behind it had originated. In Denmark, Sweden, Switzerland, Norway, Germany, and in 27 states of the USA it became a standard practise. It was not until 1935 that in America, for example, a report by the American Neurological Association recommended that sterilisation should not be carried out without the patient's consent.

It was clear to some of the Social Darwinists that some races were more desirable than others. When they considered the matter closely they decided that not only were white races more desirable than black ones, but even among Europeans there were grades of desirability. There could be a question mark over, say, the Italians. Nordic Whites, later identified as Aryans, were obviously out of the top drawer. In the hands of the totalitarian regimes of the inter-war years, the idea of preventing undesirables from perpetuating their kind was combined with absolute power to define who was and who was not desirable. In the Soviet Union undesirability was more often defined by social attitude rather than by race, but the propensity to publish criticisms of the Government could still be diagnosed as a clear symptom of mental dysfunction, and those who contracted the condition could be detained for life for their own "good" and society's. In Nazi Germany the logical way of purifying the Aryan strain was racial cleansing, leading to the Final Solution.

It was obvious that just as the Devil can quote scripture for his purpose, selected quotations from Darwin could be used in support of very different political attitudes. But long before that, even before the First World War, the intellectual Left was pursuing a campaign against Darwin's ideas on even more basic grounds. They claimed that his one great contribution to evolutionary thinking—Natural Selection—had been a mistake.

> **Summary** *Darwin and Marx exchanged polite messages but were never on the same wavelength. For decades after Darwin's death, his theory of natural selection was being co-opted by the Right for purposes he would never have condoned.*

...the images of the blacksmith and the giraffe.

Chapter 4
Lamarckists

If it could be proved that the whole universe had been produced by such Selection, only fools and rascals could bear to live.

— George Bernard Shaw

Darwin did not originate the theory of evolution. It had been around in one form or another for a long time. His own grandfather Erasmus had written at length about it—sometimes in verse—and Darwin himself as a young man had read and admired the evolutionary ideas of the pioneering French biologist Lamarck.

Lamarck not only believed in the transmutation of species, he believed that species evolved as a result of their own efforts. He evoked two images which are constantly revived to explain how his views differed from Darwin's: the images of the blacksmith and the giraffe. Lamarck noted that through persistent striving, the blacksmith is able to alter his own physique by acquiring powerful muscles. He suggested that in the same way a giraffe by persistently stretching its neck to reach higher branches could make its neck fractionally longer. He believed that changes in the blacksmith's biceps and the giraffe's neck would be passed on to the next generation and in time they would accumulate until they give rise to a new and improved species.

Lamarck was an aristocrat who at one time enjoyed royal patronage as curator of the Royal Herbarium, but when the Revolution came he was in no danger of the guillotine. His writings were particularly admired by Jean-Paul Marat, because they were non-religious and reinforced Marat's conviction that people can, if they only try hard enough, change the world, and remould it nearer to their heart's desire. So Lamarck's scientific career proceeded unhindered. He became a scientific adviser to the New Republic and masterminded its creation of a National Museum of Natural History. He was loathed by the leaders of opinion in England as a craven traitor to his class, as well as on account of the "abominable trash vomited forth by Lamarck and his disciples."

Charles Darwin (and, independently, Alfred Russell Wallace) later proposed a different explanation of why the giraffe's neck grew longer. It was not because any particular giraffe was stronger-willed than others, but because the giraffes with slightly shorter necks would have less to eat. In a bad season they would be more likely to die early, the females would produce less milk, the offspring would be less well nourished and lose out in the competition for scarce resources. So the longer-necked giraffes would outbreed the others. This process was called Natural Selection. The key to it was death—the death of the least fit in each generation.

Darwin did not believe in progress. He could and did write lyrically about the miraculous diversity of life on earth, including many creatures "most beautiful and most wonderful." But he knew that it also contained many things most ugly and disgusting and stupid. ("What a book a devil's chaplain might write on the clumsy, wasteful, blundering, low and horribly cruel works of Nature!") He was under no illusion that nature was inevitably predisposed to favour the bright and the beautiful.

He never quite rejected Lamarck's belief that characteristics acquired during a lifetime could be passed on to future generations; but the thought did not inspire him with the same optimism. He felt that natural selection exerted a far more powerful influence over how species evolve. In the photographs taken of him late in life, the expression in his eyes may partly reflect the pain he had suffered during years of ill health. But it is certainly not the expression of a man who believed he had brought glad tidings to the world. His message contained no promise that the future must be better than the past.

The version of Darwinism that Huxley publicised had been subjected to a certain amount of what would nowadays be called spin-doctoring. Among many of the ragged-trousered philosophers in his evening classes, the name of Red Lamarck was still revered as the patron saint of optimism and progress, and he had no wish to antagonise them. But he had other audiences for whom the name would spark off aversion or derision. Lamarck was by then yesterday's man. After the assassination of Marat he had been attacked by Marat's political enemies and his own scientific rivals and stripped of office; he died poor and alone and blind, without honour even in his own country. The only British academic who had ever defended him—Robert Grant at UCL—had become doddering and passé, the reverse of the keen modern image that Huxley was trying to promote. Huxley compromised by concentrating in his public lectures on the basic message of Darwinism, our close relationship with the animals, without dwelling much on the mechanism of evolutionary change. His message was accompanied by a tone of voice and a body language more appropriate to Darwin's French predecessor. There was after all no reason to think that Lamarck was going to rise from his grave.

As it turned out, it had been far too soon to write him off. His ghost continued to walk all over the evolutionary controversies of the twentieth century. One of his most voluble champions, George Bernard Shaw, declared in 1921 that while at first the Darwinian process of natural selection seems simple and intellectually satisfying, its implications are emotionally very uncomfortable. Shaw was a socialist, and one basic tenet lies at the heart of all brands of left-wing philosophy. It is the belief that while there are many things in the world and particularly in human society that we deplore, it lies within our power to try to change them. That is the proposition that in Shaw's view was under threat from Darwinism.

"When its whole significance dawns on you," he wrote, "your heart sinks into a heap of sand within you. There is a hideous fatalism about it, a ghastly and damnable reduction of beauty and intelligence, of strength and purpose, of honour and aspiration to such casually picturesque changes as an avalanche may make in a mountain landscape or a railway accident in a human figure." It works "by blindly starving and murdering everything that is not lucky enough to survive in the universal struggle for hogwash."

Shaw's success as a dramatist had given him an influential platform on the stage, and he proceeded to write a series of no less than five plays—a "Metabiological Pentateuch" entitled *Back to Methuselah*, inspired by the Lamarckian will to improve, which he called the Life Force. There must have been a time when London playgoers as a class were more or less solidly Lamarckist, though a few may have felt that they had been told more about this issue than they really wanted to know.

Shaw was not the first nor the last writer to enter this arena. The novelist Samuel Butler, who wrote *The Way of All Flesh*, was a fervent believer in evolution but he had attacked Darwin for his advocacy of natural selection as the way in which evolution worked. Shaw felt that Butler had gone too far when he "even attacked Darwin's personal character, unable to bear the fact that the author of so abhorrent a doctrine was an amiable and upright man." (Shaw himself was content to describe Darwin as "an intelligent and industrious pigeon-fancier," a description which he claimed was denounced as a blasphemous levity.) And later Arthur Koestler, author of *Darkness at Noon*, was deflected from literature and politics to immerse himself in the Lamarck controversy and write books about it.

To most scientists this was an inexplicable phenomenon. Why were these people trying to get in on the act? One of my favourite scientists, Peter Medawar, expressed utter astonishment that this was a topic "upon which literary people for some reason felt themselves entitled to express an opinion."

All that was needed to prove Lamarckism true was a single well-attested example of an acquired characteristic being passed on to the next generation. That would settle it. However, the absence of such an example settled nothing. It only meant that the search was still in progress. The Lamarckists were travelling hopefully but failing to arrive.

The strength of their conviction led in one or two instances to fraudulent claims, but more often to "nothing more culpable than self deception." Meanwhile their opponents consolidated their ground. The German biologist August Weismann asserted as an immutable law that the cells of the "germ plasm" (sperm cells and egg cells) could pass information and instructions to other cells of the body, but nothing could pass back into them, and hence into the next generation. When he proposed it, it was little more than a hunch, but advances in scientific knowledge seemed to confirm it. Finally, rephrased as "nothing can pass from proteins back into nucleic acids," it was proclaimed by Francis Crick to be "the central dogma" of orthodox biological thinking.

The protests voiced by the literati had had little effect on the outcome of the argument, but there was more powerful pressure from another non-scientific source: the political theorists. Marxist philosophers passionately wished to believe that Man had risen by his own efforts and that manipulating the environment could change any living thing. By the time that Stalin had imposed his own totalitarian style of Communism on the Soviet Union, anything that he wished to believe was proclaimed to be the truth. When an agricultural plant breeder called T. D. Lysenko managed to ingratiate himself with Stalin, he was entrusted with the task of expounding a specific Stalinist approach to biological dogma, and no other version was taught in Russian schools for a generation.

At a session of the Lenin Academy of Agricultural Sciences Lysenko proclaimed that "Darwin's theory, though unquestionably materialist in its main features, is not free from some serious errors." Fortunately, he said, Marxist analysis had succeeded in revealing those errors. Biologists who failed to understand that were named and shamed. At the end of the meeting a resolution was passed denouncing "the reactionary biologists Weissman, Mendel, and Morgan," and proclaiming that "The new characters which plants and animals acquire under the influence of their conditions of life can be transmitted by inheritance."

Lysenko's experiments on Soviet agriculture were based on the assumption that crops could be gradually conditioned to thrive in colder latitudes, and would pass on that acquired characteristic. Richard Dawkins in *The Blind Watchmaker* described the aftermath. "Incalculable damage was done to Soviet agriculture. Many distinguished Soviet geneticists were banished, exiled, or imprisoned. For example, N. I. Vavilov, a geneticist of worldwide reputation, died of malnutrition in a windowless prison cell after a prolonged trial at on ludicrously trumped up charges such as 'spying for the British.'"

For people like John Maynard Smith who at that time was both a scientist and a Marxist, it posed a crucial dilemma. He once recalled, in the course of a book review, "I couldn't have things both ways. I even put in six months trying to do a Lamarckian experiment to demonstrate inheritance of an acquired characteristic, but of course it failed."

In effect that war is over. It is true that the central dogma turned out to be less inviolate than was believed at one time. Experiments with bacteria conducted at Harvard by R. B. and R. M. Hamilton indicated that mutations had been induced in some strains by nutrient deprivation, under conditions normally associated with selection pressure, and passed on to succeeding generations. It was discovered that a substance called reverse transcriptase can retranslate RNA into DNA, and retroviruses can use this route to place new material into chromosomal DNA. There are occasional reports of epigenetic changes observed in, for example, a toadflax plant—changes not linked to alterations in DNA sequences, but nevertheless inherited by subsequent generations. So it is now (just barely) permissible to suggest that the central dogma may have been slightly too dogmatic.

However, the non-scientists who took part in the debate were not thinking in terms of bacteria. They were thinking about people. In an organism of the size and complexity of *Homo sapiens*, evolution through will power, by means of the Shavian Life Force, does not work. You may by diligence and determination increase your physical or mental capacities considerably, but the increased capacity will never be passed on to your children in your genes. Some people find that depressing, and mourn that the world would be a better place to live in if Lamarck had been right.

I can't help feeling that they haven't thought it through. It is true there are men and women who feel that they have made the most of every last ounce of their DNA potential—stretched it to its limits and maybe extended them. Naturally they think it hard that their children should have to start again at square one. Perhaps they are over-represented among the types of people who argue about evolution. But behind this attitude there is an unstated premise—namely, that Lamarckism would only entail passing on desirable acquired characteristics, and would magically screen out undesirable ones. Such a mechanism is very hard to envisage.

In the course of its development from embryo to adult, an individual organism incurs far more numerous and varied kinds of damage than were ever imprinted in its genes. If all these were passed on to the genes, and had to be slowly and separately phased out again by natural selection, our fitness to survive would rapidly deteriorate.

There are millions of people around the world who look back on their lives and see how they were kept ignorant by neglect, made fearful and bitter by abuse, their bodies and minds stunted or damaged by malnutrition, oppression, infections, parasites, drugs, landmines. . .For such people, if those acquired characteristics were all liable to be passed on, every childbirth would be anticipated with dread. The window of hope for them is that every conception is a brand new start. Just possibly, over the next ten years, the monsoon will not fail; the cease-fire will hold; someone will dig the well and build the school, and then the child will become not as they are, but as they might have been.

So, unlike Shaw, I can get along without Lamarck and still not feel my heart sinking into a heap of sand.

> **Summary** *Many people on the Left tried hard to prove that acquired characteristics can be inherited as Lamarck believed. They failed. What we do in our lifetime does powerfully affect the prospects of our successors, but this influence is not exerted via the DNA.*

They will follow anything in front of them that moves.

Chapter 5

The Twig Is Bent

'Tis education forms the common mind. Just as the twig is bent, the tree's inclined.

— Alexander Pope

Pope's simile was drawn from horticulture. But in the past, the intuitive folk wisdom about human behaviour was more often drawn from stock rearing, and suggested that the way an animal conducted itself was "all in the blood." Any fool could see that cows are born to behave in one way and horses in another way. As for humans, they were living in a stratified society where there was virtually no mobility between the classes. Peasants begat peasants and gentlemen begat gentlemen, as reliably as sheep begat sheep. It seemed as clear as day that a young lord must have inherited his father's arrogance, as well as his hair colour and his estate.

R. C. Lewontin has illustrated the strength of this belief by referring his readers not to the *Origin of Species*, but to a book written 10 years earlier by a different Charles. In the novel *Oliver Twist*, Dickens described a boy whose mother had died in giving birth to him so that he was brought up in a workhouse, among the orphaned or abandoned children of the dregs of society. These boys were underfed, neglected and abused by hard-faced officials. They grew up rough and ignorant, "rolling around on the floor," the weaker ones broken in spirit and the stronger ones coarse, crafty and aggressive. But Oliver blossomed in that stony ground. He developed into an honest, brave, gentle and polite little boy, instinctively speaking—unlike any of his fellows—good grammatical English. Dickens's readers like Dickens himself would have envisaged no other possible outcome, since Oliver's mother was a lady.

Darwin's work made little impact on that perception. If anything it reinforced it, since natural selection is exclusively concerned with heredity. Within ten years of his death, however, two very different men had begun to study how behaviour is affected by events that occur after birth. They were Sigmund Freud in Vienna, and I. P. Pavlov in St. Petersburg.

Freud is a classic example of how some ideas can spill out of the textbooks and into the public consciousness almost as fast as they are published. For a couple of generations, several of his concepts, such as the subconscious mind, infantile sexuality, the Oedipus complex, and the meaning of dreams, were quoted wherever educated people entered into discussions about human nature. There were several reasons for this. For one thing the ideas were promoted not as abstract speculations about the nature of humankind but as therapy. To Macbeth's poignant question: "Canst thou not minister to a mind diseased? Pluck from the memory a rooted sorrow?" Freud in effect answered: "Yes, I can." It is debatable whether any of the patients treated by Freud recovered more readily or permanently than those treated by other means or left untreated. But the very fact that someone was claiming to have a brand new way of coping with mental illness gave grounds for hope.

Freud also suggested a useful new way of thinking about the psyche. The word "subconscious" depicted the mind as a layered organ with the reasonable layer on top and a mess of more primitive stuff underneath it. For those who have experienced the feeling of "I can't think what came over me!" (i.e. all of us) it is slightly less spooky to think of something seeping up from the unconscious than to be told (as our ancestors would once have been) that we were possessed by an evil spirit. Freud was right too, in judging that in the society where he lived and moved, there was far too much secrecy and shame and taboo attached to the subject of sex.

But Freud does not fit very naturally into a book about evolutionary scientists. Many people question whether he was a scientist at all in any recognisable sense of the word. For example, his pronouncements about the stages of a baby's consciousness, and how it feels about its faeces and its father, were not based on any observation of babies. They were based on observing the way some dysfunctional adults behaved, and constructing a fable about how they might have got that way. Attempts to take his ideas literally began to falter with the passage of time. For one thing, in blaming so much stress and malfunction on the suppression of sexual information, he led many to believe that if ever the time came when sexual appetites might be freely admitted to and discussed, human relationships would automatically become far more relaxed and serene. That expectation has fallen flat on its face.

I. P. Pavlov was a very different proposition and his ideas got a very different reception. He was a brilliant surgeon who made discoveries in the fields of cardiac physiology and gastric secretions that earned him a Nobel Prize. In his investigation of animal behaviour he was interested by the basic scientific questions: What exactly is going on here and how does it work? And he employed impeccably scientific methods to arrive at the answers.

Typical of the Pavlovian approach are the experiments that teach a rat to find the quickest away through a maze.

It learns the task very readily because it is born with two instincts. One is an appetite for exploring its environment: we may call that "curiosity." Pavlov called it "latent learning." The second is an instinct to repeat actions which lead to a satisfactory experience and avoid actions that lead to an unpleasant one. A rat in a maze which finds that the second turning on the left leads to a dead end will on future journeys avoid the second turning on the left. Folk wisdom knows all about that one too. We may call it "trial and error" or "once bitten twice shy." Pavlov called it a "conditioned reflex." By a suitable system of rewards and punishments a rat or a pigeon or a squirrel or a human being can be trained to perform truly remarkable feats.

In theory this general approach, which became known as behaviourism, would seem more congenial to left-wing thinkers than attributing everything to heredity. But some of the behaviourists carried their convictions to extremes where it was hard to follow them. Some of Pavlov's followers suggested that when we are born our minds are like blank slates for experience to write on. One of them, J. B. Watson, went so far as to assert that "there is no such thing as instinct." Still today, "blank-slater" is the epithet of choice to hurl at left-wing theorists, just as "genetic determinist" is used to discredit the Right.

The ideas of the behaviourists were pounced on by naysayers galore. People did not like being told that their brains were, or ever had been, blank slates. Shaw became a spokesman for the protestors, moved chiefly by the fact that the research involved experiments on captive animals, and he was a passionate anti-vivisectionist. Some of the experiments did indeed involve cruelty and would not be allowed today. But most of Pavlov's "conditioning" of rats and pigeons was carried out in the same way as a dog is trained to herd sheep or become a guide dog for the blind.

Shaw however poured ridicule on the whole approach, arguing that at best it only "discovered" things that everybody knew already, and no good could ever come of it. Why measure a dog's saliva to prove that it can associate a given sound with the prospect of food? Everybody knows that a hungry man's mouth may water when he hears the dinner bell.

The behaviourist approach is often presented in the worst possible light. A favourite target is the somewhat eccentric professor B. F. Skinner and the notorious Skinner Box he designed for his baby daughter. It was widely represented as a kind of sterile coffin in which she was confined like a battery hen to protect her against infection. It was apparently a purpose-planned air-conditioned environment in which she could when desired be left to her own devices and come to no harm—a kind of glorified archetypal play-pen. And what Skinner claimed to have discovered through his animal experiments was that if you want to influence anybody's behaviour, the carrot is more effective than the stick, as well as more humane.

It was not true that the early investigations into animal behaviour only discovered things that were already obvious to everybody. It was surprising, for instance, to learn that among primates two of the most apparently hard-wired instincts of all—the mating instinct and the maternal instinct—are not really hard-wired, but partially imitated. A male monkey brought up in isolation doesn't know how to copulate, and a female one brought up in isolation takes no interest in her babies.

Hostility to the behaviourists continued for a long time. By the mid-century when totalitarian governments came into power, behaviourist scientists were sometimes accused of having helped them to get into power by inventing the technique of "brainwashing." Sadly, tyrants and priestly Inquisitors down the ages have never been short of ways of producing instant converts to their own particular creed. That charge was a bit like blaming Michael Faraday for the electric chairs on Death Row.

In the Netherlands in the 1930s, Niko Tinbergen began looking at animal behaviour from the opposite end of the spectrum. He studied birds and animals in the wild, and concentrated not on their learned behaviour, but on the kinds of behaviour they did not need to learn because they were instinctive. For this purpose, the favourite subjects of research were often birds—the gull, and the goose. It was not hard, for example, to prove that goslings do not follow their mothers because they have learned by trial and error that it is a sensible plan. It is an instinct. At a given stage of their development they will start to follow anything in front of them that moves and calls back to them when they call. In the 1930s, that moving object was frequently not their mother but Konrad Lorenz, who collaborated with Tinbergen and was responsible for introducing this new field of research to a wider public. It became known as ethology.

Where behavioural training had often made its subject species look surprisingly clever, ethology often made them appear surprisingly stupid. A goose for example seems a wise and concerned mother when she retrieves an egg that has rolled out of the nest, or welcomes her little ones under her wing. But she will just as carefully retrieve a cardboard cube if it is roughly the size and colour of an egg, and will tuck a stuffed stoat under her wing if it has been equipped with the ability to cheep like a gosling.

Outside the academic world, most people were much happier with this kind of analysis of what makes us tick. It is not entirely obvious why they took it to their hearts so readily. If it had not been flattering to be told that we learn in the same way that animals learn, why were we happier to hear that we are driven by the same unconditioned instincts as animals are? It may well have been partly due to the charismatic writing skills of Konrad Lorenz. His best-selling books *King Solomon's Ring* and *On Aggression* featured an endless stream of amusing animal stories—like the one about the jackdaw who fell in love with him and showed her devotion by stuffing minced worms into his ear, for the lack of any other suitably sized orifice.

For whatever reason, the insights derived from ethology were readily accepted. Terms derived from ethological studies soon became common currency, and began to displace some of the Freudian patois of complexes and transferences. People seem to feel that they understand human behaviour better (especially other people's behaviour) if they can describe it in ethological terms. Thus, if somebody tries to upstage us, we tell ourselves that they are mindlessly trying to enforce a pecking order, like a stupid chicken. If a sister offers to baby-sit, it is kind of her of course, but it is really the least she could do since she must be motivated by kin selection. If rivals are sucking up to the boss, they are obviously grooming him with flattery, or demonstrating appeasement behaviour to the pack's dominant male. If there is an argument over where cars are parked, somebody's territorial instincts must be getting out of hand; he is trying to claim an increased area of personal space, like any touchy baboon.

The difference between the behaviourists' approach and the

ethologists' approach has been a recurring theme in evolutionary discussions ever since Darwin. It is the old question of Heredity versus Environment, or Nature versus Nurture. There should of course be nothing "versus" about it, since everyone agrees that all living creatures are the result of the interaction of heredity and environment, and you can't have one without the other. There is however an inveterate tendency for individuals or groups of people to protest from time to time that the importance of one or other of those factors has been grossly exaggerated.

Which side you are on may depend partly on the job you're doing. Most people like to assume that what they're doing is important. It is a healthy assumption. So if you ask a doctor to account for the drop in the death rate in Britain in the 20th century, he will point to the advances in medicine—antiseptics, vaccination, improved surgical procedures, antibiotics. If you asked a social reformer the same question he would attribute it to improved public sanitation, welfare provisions, slum clearances, health education, and Clean Air Acts. Both would be correct.

The other factor affecting the perception of bias is political. People who are on balance contented with the way human society is organised tend to lay stress on heredity. They predict that pipe-dreams of a more egalitarian arrangement will always be frustrated by the immutable brute facts about human nature. Those on the Left assume that most of the existing inequality has been imposed rather than inherited, and could be drastically reduced or eliminated by altering the social environment.

In the last few decades, the tide of opinion has set strongly in favour of the hereditarians. One reason may have been the collapse of the hopes that had been invested in the creation of states that set out to be socialist, and a consequent loss of morale on the Left. The human failings of greed, ambition, and aggression are not so easily eliminated by changing the social landscape. At the same time the rise of global capitalism has led to a new enthusiasm among entrepreneurs and speculators for exploring the mercantile interpretations of biological theories, leading to a kind of consilience between genetics and economics. If greed is hard-wired in our chromosomes it must be adaptive, so why not learn to love it and remove all attempts to restrict its operation?

The third reason was the advance in technology, which ensured that most of the exciting new biological discoveries, and most of the jobs, and most of the money available for investment in research, were all concentrated on genetics. No one denies that nurture has a part to play, but if you spend your working life thinking about genes, they will inevitably come to occupy a pivotal place in your thinking. "There is nothing," as the shoemakers used to say, "like leather."

Whenever such a tide is flowing, there is some danger that any hypothesis which appears to conform to it will be accepted uncritically, without being properly examined. I propose to show that this is indeed happening, and that some unsafe ideas have gained a foothold in the conventional wisdom as a result.

> **Summary** *In the twentieth century, different methods of researching led to conflicting interpretations. Behaviourists concluded that learning was the main factor in determining animal behaviour. Ethologists concentrated on those aspects of it that were innate. The tension between the Nature and Nurture factions has since taken different forms but has never quite gone away.*

"This can't be right."

Chapter 6
Thicker than Water

At times I was sure I saw something that others had not seen.

— W. D. Hamilton

Ever since Darwin's day, evolutionists have been bothered by the "problem of altruism." If all of life is a struggle for hogwash, and only the fittest survive, why do people ever co-operate with others, or give them a helping hand, instead of competing with them? By the middle of the last century that had come to be regarded as one of the central theoretical problems of human evolution.

One theory was that they were co-operating in the interests of the tribe to which they belonged. It suggested that when groups of animals compete, say, to occupy a particularly desirable piece of territory, the group which co-operated would prevail over the group in which individuals wasted time and energy squabbling among themselves. So such behaviour would be selected for. It sounded reasonable, and not only to scientists. "All for one and one for all" was not a motto that everybody lived up to, but everybody understood it. Most of them could remember instances of people "sticking together" or "rallying around" in times of crisis.

Those on the Left who dream of building a peaceful and co-operative society would like to believe that this tendency to co-operate for the common good is deeply rooted in human nature. Some people are ambiguous about it. Margaret Thatcher, for example, when talking about income tax and public expenditure, famously declared that "There is no such thing as society." Yet when preparing to reclaim the Falklands, she strongly favoured solidarity, fidelity, obedience, and the willingness to die for that non-existent society.

The Scottish ecologist V. C. Wynne-Edwards was convinced that there was an inherited instinct involved, and he christened the process "group selection." If it was indeed an instinct, there ought to be examples of it in other vertebrates—but he was not notably successful in finding them. He wrote about one species which he had studied in detail—the red grouse. He suggested that when they congregate together in a flock it is in order to monitor their population density; so that if they are too thick on the ground, individuals can voluntarily restrict their fertility for the common good; and he extended the idea to explain other phenomena like birdsong and the movements of zooplankton. But the concept of voluntarily restricting fertility is very hard to substantiate. You can prove that a period of overcrowding is in fact followed by a period of lower birth-rates, but that may simply be because food has already become a relatively scarce resource. Hunger is a very potent fertility reducer.

Other believers in group selection preferred to cite those archetypal altruists and co-operators, the social insects. But not everyone relished the idea of comparing *Homo sapiens* to bees and termites. An American called George Williams who studied at the University of Chicago was thoroughly turned off by it. He relieved his feelings by writing a book attacking the whole concept of group selection, which was almost instantly acclaimed as a landmark in Darwinian thinking.

Williams declared that group selection was an untenable idea. He pointed out that if a group of animals evolved in which all were programmed to value the interests of the community above their individual interests, it would need only one defector to arise in its ranks to cause the whole system to collapse. The defector would reap the benefits of co-operation without paying his share of the cost and would therefore be the best fitted to outcompete everyone else in the group. His progeny would multiply and spread their inheritance through the group until there were no co-operators left. It was a robust argument, cogently expressed, and from that point on, group selection was widely reckoned to be a dead duck.

Not everyone agreed. According to Elliott Sobers and D. S. Wilson, Williams's arguments against group selection were not conclusive and the accounts of its death were much exaggerated. "Williams's rejection of it," they wrote, "was celebrated as a scientific advance comparable to the rejection of Lamarckism, that allowed biologists to close the book on one sector of possibilities and concentrate their attention elsewhere." They report that some terse advice was offered to students by one "very distinguished evolutionary biologist: 'There are three ideas that you do not evoke in biology—Lamarckism, the phlogiston theory, and group selection.'"

So there were some people who saw group selection as an idea tainted by wishful thinking and would fight to the death against giving it houseroom, and a smaller number prepared to defend it. As often happens, the longer the argument continued, the nearer it came to revealing itself as largely a disagreement over terminology. Nowadays the term "neo-group selection" is sometimes used to skirt around the problem, and younger students often seem surprised to learn that it was ever a big deal, one way or the other.

For an Englishman called William Hamilton it was a *very* big deal. He has been described as being "allergic" to group selection, and his forceful rejection of it threw the problem of altruism into high relief. If people were not nice to each other because of group selection, why *were* they sometimes nice to each other? Was their apparent niceness all hypocrisy and self deception? The problem came to obsess some of the most creative minds of the day, just as Darwin had been obsessed by the bewildering number of species in the world, and Mendel by the problem of throwbacks in breeding stock.

Hamilton had studied the social insects and was very keen to explore a possible connection between altruism and genes. Having no backer or adviser or Ph.D., he found it hard to interest any university in his proposition. The biology departments were interested in genes but not altruism. In the end he applied to the London School of Economics, which was interested in altruism but not (at that time) in genes. However it offered him a scholarship and access to libraries and let him get on with it.

Hamilton had noted the unusual way in which bees reproduce which results in a worker bee being more closely related to her sisters than she would be to her own children if she had any. By remaining sterile and helping the Queen to produce more sisters, she can ensure the production of more copies of her genes than if she began laying eggs of her own. So if you think about what is in the selfish interest of the genes, instead of what is in the selfish interest of the organism, this apparent "altruism" makes perfect sense. This way of thinking was not *entirely* new. J. B. S. Haldane had verbally speculated along those lines in 1955, saying that natural selection should cause him to be willing to lay down his life for two brothers or eight cousins—but it had been a jocular off-the-cuff remark so no one had taken it very seriously.

Hamilton's 1964 paper was neither jocular nor casual. In fact it was in danger of being ignored for the opposite reason. It was long, detailed, laboriously argued and punctuated with pages of extremely complex mathematics which very few biologists at that date were able to evaluate. Some didn't even try, like the zoologist Sir Solly Zuckerman, who admitted that when reading a scientific paper featuring mathematical formulas—"I hum them." Nevertheless, when Hamilton had followed the advice of John Maynard Smith to submit the paper in two separate parts, he succeeded in getting it published.

Within a few years a surprising fact came to light. It was announced that Hamilton's paper, judging by the number of times it had been cited in papers by other scientists, was the most influential one *that had ever appeared* in a professional journal. It is true that many—perhaps most—of the people who cited it had not actually read it. (Hamilton himself pointed this out rather sardonically: one or two of the earliest commentators on it had misquoted the title, and the hundreds who came after them copied the misquotation.) But the basic message they carried away was simple and revolutionary.

Hamilton transformed the Darwinian idea of the survival of the fittest by introducing the new concept of *"inclusive* fitness." This measured an individual's ability not merely to survive to puberty, but to produce offspring healthy enough to survive and reproduce in their turn. Natural selection was interested in the grandchildren. And the underlying reason for this was that evolution is exclusively concerned with perpetuating our genes, and not interested in us except as the carriers of those genes.

E. O. Wilson gave a vivid account of how he opened Hamilton's paper on a train departing from Boston and began to read, thinking "This can't be right." By the time the train arrived at Miami he had experienced a conversion like that of St. Paul on the road to Damascus: "I gave up. I was a convert." John Maynard Smith's reaction was equally profound. He felt as Huxley had felt about Darwin: "Now why didn't I think of that!"

This was a crucial turning point in the thinking of biologists. Hamilton's paper affected the standing of biology within the scientific pecking order, because it was seriously mathematical. Biology had hitherto been relatively short on maths. Darwin's *Origin of Species* is virtually devoid of it. It is just words and pictures, with no tables or formulae or equations or pie-charts or graphs, and only one diagram. Mendel had indeed introduced a numerical element but it was a fairly simple one. Hamilton's was far from simple, but even for those of his colleagues who couldn't follow it, it was good for their morale. And the exciting part was that Hamilton was not doing sums about anatomy (measuring things like pulse rate and body temperature and blood pressure were familiar enough) he was doing sums about *behaviour*. That must by implication include human behaviour. And who could tell where that might lead?

It also caused another shift in the delicate balance of the Nature/Nurture continuum. From that point on, genetics became the biological growth point, the place where the action was. It was full of promise. It attracted funds. It was constantly spewing out answerable questions, and crying out for students to come and work on them. Some of their colleagues, working on other valid approaches to life on earth, had begun to feel marginalised and were not too happy about it, but the process so far shows no sign of slowing down.

Hamilton had solved—or very nearly solved—the problem of altruism. He named the gene-based mechanism "kin selection." His conclusions have been exhaustively checked and applied to a very wide range of species on land and sea, and they seem to work in all cases. His theory confirmed the intuitive folk wisdom which had always held that blood is thicker than water.

Like Darwin, although he had introduced a new way of thinking about life on earth, he was under no illusion that he was bringing glad tidings. He had hoped to show that altruism is as deeply rooted in genetics as aggression is, which in itself would have been a message to lift our hearts. But in following truth wherever it led, he reached a bleaker conclusion. If he was right, then it was possible that human altruism had little to do with love or kindness or even with humanity. It was merely a disguised form of the drive for self-preservation in our genes, dictated to us as to every other species by our DNA. Describing his own feelings about it, he wrote "A scientist or philosopher with a programme of such heresy has to be tough if he or she is to communicate it, and while doing so and for long after, must endure the tortures of Orestes."

When George Price first read Hamilton's 1964 paper he actively disliked it and set about acquiring enough scientific and mathematical skill to refute it. Hamilton later recalled: "The paper seemed to have a profound effect on him. He set to work to try to understand genetics and to verify what I had done. Once he had convinced himself that something at least close to what I claimed was true, he became very depressed."

Price then entered the field himself, and his contributions to it were original and brilliant. He collaborated with Hamilton. His ideas on the evolution of animal fighting helped to inspire a seminal paper by John Maynard Smith. But he never succeeded in reconciling what his head told him was true with what his heart felt about it. He converted to a very literal form of Christianity, selling all he had and giving the proceeds to the poor. He ended up living alone in a squat in Euston, offering food and shelter and sympathy to vagrants and alcoholics, and in the winter of 1974 he committed suicide.

That does not mean that the "pain" felt by Hamilton, and even more acutely by Price, was an inevitable result of their work on genetics. Humans are complex organisms and if those two had worked in another sphere, they might have found something else to be depressed about. Others were more resilient. John Maynard Smith, while sharing the feeling that gene selection involved making a choice between truth and optimism, was more stoical. He decided he could live without the optimism. Stephen Jay Gould commented that we already live with several unpleasant biological truths, including the fact that we are all going to die. If genetic determinism turned out to be true, we would learn to live with that as well. No doubt some people who had always held a jaundiced view of the human race may even have felt some satisfaction in being able to say in effect "I told you so."

As long as altruism had remained unexplained, there was just a chance that, in humans at least, some other factor was involved—not just another variation on the hogwash theme. But Hamilton had explained away a great swathe of "altruistic" behaviour by reducing it to kin selection, a genetic predisposition to behave well not only towards offspring but to others more distantly related by blood. Most animal populations which act co-operatively are interconnected by a network of blood relationships, and kin selection made perfect sense of their behaviour. Problem solved?

Virtually solved, but not yet watertight. People are occasionally nice to others who are not connected to them by ties of blood. It is not unknown for someone to offer to see an old lady across the road, even if she is not his Auntie Ida. He is not doing it for himself; he is not doing it as a favour to his genes; he is not doing it for England. Why is he doing it? It was one of those minor loose ends. Sooner or later somebody would have to come along and tidy it up. And the sooner the better.

> **Summary** *Scientists who concentrated on the inherited component of animal behaviour faced the problem of how altruism could ever have been adaptive. The search for a solution led George Hamilton to identify the gene, rather than the organism, as the unit which Natural Selection was interested in preserving.*

...as out of place as at a Saturday night poker game.

Chapter 7
Bread upon the Waters

In the early 1970s the whole field of animal behaviour was "stampeding"—to use Hamilton's expression—in the direction of the new ideas of kin selection and reciprocal altruism.

— Ullica Segerstråle

Robert Trivers was a brilliant young student at Harvard who was inspired by Hamilton's new slant on natural selection. He had no difficulty at all with the mathematics. He even detected a minor flaw in it, and though Hamilton himself had become aware of his error no one else had noticed it. Trivers was a better communicator than Hamilton (which was not difficult). He expounded the implications of kin selection to his mentor Irven de Vore, who instantly perceived that "this may be hot!" and set about organising seminars so that other students could also be exposed to it. It was not long before de Vore concluded that "you have to divide the world into pre- and post-Bill Hamilton." It was largely because of Trivers that Hamilton's work was, for a period, better known in America than in his own country.

Trivers was after the same Holy Grail that Hamilton had pursued—an explanation of altruism in terms of natural selection. Hamilton had begun with an open mind and had only finally been driven to a gene-centred explanation of animal behaviour. But Trivers, after reading Hamilton, treated that as his starting-point. His own aim was to tidy up the loose ends by explaining that part of human behaviour which kin selection had failed to account for—the fact that people sometimes behave unselfishly to people who are not their relatives.

He was well endowed with energy and imagination and it was not long before he arrived at a theory of his own—the suggestion that it could be adaptive, in the Darwinian sense, for one individual to inherit the practice of behaving benevolently to another (unrelated) one, if there was a probability that the favour would in some way or other be returned. In that way the individual's inclusive fitness would be increased, and the gene for such behaviour would be selected for. He called this process Reciprocal Altruism.

Nothing could sound more natural and probable. It takes place all the time. The vernacular phrase for it is "You scratch my back and I'll scratch yours." There is even a text in the Bible advising us to cast our bread upon the waters, with the reassuring promise that "thou shalt find it after many days." The language is full of sayings acknowledging that when favours are done they must be returned. "Right, I'll do it—but remember: you owe me!" or "I wouldn't ask him to do it for me, because I don't want to be obligated to him."

The idea has been widely accepted. For some strange reason the favourite image chosen to illustrate it—the equivalent of the giraffe's neck—is the suggestion that you might save a man from drowning in the hope that if he ever sees you drowning, he will return the favour. If he could not even keep himself afloat, it seems rather a long shot. But the principle behind it is clear.

The problem is that it does look very much like a learned strategy. Children would find out very soon, by trial and error, that if you are nice to other people they will be more disposed to like you, and if they like you they will be more disposed to be nice to you in return. It is not only learned by experience; it is also passed on by injunction, as in the Biblical text. We do not generally need injunctions to do what our instincts prompt us to do; there is no text instructing us: "When thou art hungry, then shalt thou eat." But Trivers wanted us not to have to learn to do it, or be told to do it. He wanted it to be in our genes, and he wanted to prove that it was.

He approached the task in a rigorously scientific manner. To show that it was a naturally inherited behaviour pattern, he supplied examples of unselfish behaviour in the natural world. For example, in some species of birds and animals a "sentinel" will alert others to the arrival of a predator, even though its call may endanger its own life by drawing the predator's attention to its presence. Cleaner fish will remove parasites from within the mouths of other fish which are big enough to swallow them whole, because they know that in return for this service their lives will be spared—and so on.

Secondly, since not all animals display this pattern of behaviour, it was necessary to draw up a list of the preconditions without which the gene for Reciprocal Altruism would not be acquired. Of course it would only arise in social species; solitary ones would not often have the opportunity to reclaim the favour. Again, it could only arise in species with enough memory and intelligence to distinguish between other individuals and remember how they had behaved in the past. For a deed to qualify as genuine reciprocity, time had to elapse between doing a favour and having it returned.

Trivers' hypothesis about Reciprocal Altruism (RA) found plenty of takers. The ground had been well prepared to receive it; it almost seemed like a continuation of kin selection by other means. The two terms, often used in conjunction, became part of the common currency of evolutionary debate. Trivers knew that the weak point in his case was exactly the same as the weak point in the old case for group selection: somebody might cheat. They might let you scratch their back and then refuse to scratch yours, so that mutual mistrust would spread and the whole system would disintegrate.

These misgivings were partly overcome by pointing out how well equipped *Homo sapiens* is to avoid that danger. The species is notoriously full of cheats and liars, but humans also have large brains and long memories and an astoundingly good talent for distinguishing one person from another. A sizeable section of our brains specialises in identifying faces, and interpreting their expressions. Even little babies can do it. And facial expression is one of the factors most likely to betray the person who is trying to pull a fast one. Perhaps then these talents were a prerequisite for indulging in reciprocity? Or perhaps after our ancestors began to indulge in reciprocity, was there a premium on genes for long memories and shrewdness? Either way, it looked like a correlation.

The mathematics of RA proved more complex than the mathematics of kinship. You can construct a family tree of your relations, but your friendships are not so easy to fit into a diagram. You are talking here not about individuals or chromosomes, you are talking about interactions between people—in other words, social strategies. It sounds quite possible to put a simple algorithm into your computer such as the Tit for Tat strategy used by reciprocators: "If you are good to me, I will be good to you; otherwise not." But if you do put that strategy into the computer, you will find that it will be promptly outcompeted by any other player who practises a strategy of "always cheat." So it would not be adaptive, and would never be coded for in the genes. It was a problem.

Fortunately it was just the kind of problem that many scientists take pleasure in. There were people who had spent years thinking about strategies, and devising thought experiments about imaginary players who are all motivated by the desire to put one over on each other, and are allotted different strategies with which to attempt it. The problem posed by Trivers's theory looked clearly soluble, especially with the aid of the latest gadgets, and it was just difficult enough to be challenging. A large number of people were attracted to it like moths to a flame.

The classic example of these strategy games was called Prisoner's Dilemma. For a period it became an icon of evolutionary thinking second only to the neck of the giraffe. Some of the best minds in academia were obsessed with finding a strategy-based formula that would explain why there are still a few nice guys around in the world. For the real addicts it became a threat to their day jobs and their love lives. Contestants in different time zones would phone and wake each other up to say "By George, I think I've got it!" only to perceive ten minutes later that their solution had a snag in it. They tried out strategies called Tit for Two Tats, or Naïve Prober, or Grudger or Retaliator. There was a built-in assumption that each player was hell-bent on doing down everybody else, so the less hard-nosed strategists were never given names like Grateful or Pardoner; it was assumed they were as greedy as everybody else but too thick to get the hang of the rules. They were tersely labelled Sucker. People would invent games beginning: "Suppose there are two individuals called Konrad and Niko." No one ever supposed there were two individuals called Mary and Elizabeth. They would have been as out of place there as at a Saturday night poker game.

Robert Axelrod organised a series of tournaments, and in the end an answer was arrived at. The secret was that the contestants had to play a long series of games in succession, and in the (very) long run, the nice Tit for Tat strategy would prevail. For example, the success rate of one "nasty" strategy would rise steeply for the first 150 generations, and then start to decline, "approaching extinction around generation 1000." That was felt to be only right and fitting, because evolution is a slow business, and natural selection only operates over a long series of generations.

The quest was over. Matt Ridley commented that some of the theorists had to be "dragged kicking and screaming back to the real world." There was a general feeling that RA had been validated. A test for it had been proposed which, with the aid of the Prisoner's Dilemma contingent, it had triumphantly passed. It had taken up a good deal of the time of a number of individual scientists, and everybody felt that it was time to move on. It was tacitly agreed that kin selection plus RA was an assumption that it would be henceforward safe to build on. Ridley described game theory as "an esoteric branch of mathematics that provides a strange bridge between biology and economics. The game has been central to one of the most exciting scientific discoveries of recent years—nothing less than an understanding of why people are nice to each other."

I wish I could share his excitement and his admiration of that strange and rickety bridge. I agree with him about the brilliance and originality of many of Robert Trivers's contributions to evolutionary thinking. (For example, I believe he was the first to incorporate the baby's-eye view into an account of the human life-cycle.)

Many if not most of his ideas have been accepted into the conventional wisdom. But personally I have grave doubts about Reciprocal Altruism. I would like to argue that people who blindly followed Trivers in that particular assumption were "stampeding" up a cul-de-sac. This raises some basic questions about how Trivers's approach differed from Hamilton's.

In the first place Hamilton's kin selection was talking about a very ancient and deep rooted phenomenon which he found wherever he looked for it—bees, termites, beetles, birds, baboons, chimps, lions, dogs, fish, mice, hyenas, langurs, whelks, and people, and plants. The one species he felt least inspired to write about was *Homo sapiens*. That of course was the one most people wanted him to talk about. He recalled how they bribed him to talk about it with such temptations as "reputation, money (honoraria) and travel to exciting places." But he was not proud of his contributions to such symposia and attributed their unsatisfactory nature to "haste, lack of focus, and lack of anything new to say."

Trivers by contrast appears to have started at the human end. For humans, Reciprocal Altruism obviously works; it makes sense. But if Trivers wanted to demonstrate that this was not learned behaviour but an inherited instinct, it was essential to find examples of it in non-human species.

He did his best. And after his theory had become a scientific front runner, others also did their best. Field workers were sent out into the wide world to collect hundreds of examples of Reciprocal Altruism as Hamilton had collected hundreds of examples of kin selection. They came back empty-handed except for one trophy—a hypothetical interpretation of the activities of a vampire bat. That will be taken as a wildly unfair statement unless we take a closer look at some of the other proffered examples of Reciprocal Altruism.

1. Cleaner-fish. These are fish that remove parasites from larger fish. The theory suggests that they are being altruistic to the host fish by cleaning it, and the host altruistically returns the favour by not eating them when it has the chance. In fact the cleaners do not eat the parasites out of kindness or in the hope of getting some kindness in return. They eat them because the parasites are their staple diet. A different cleaner-fish obligingly removes parasites from broad-leaved kelp in exactly the same way, without expecting the seaweed to repay the favour.

2. Sentinel behaviour. This refers to a bird or animal which keeps a lookout and raises the alarm when a predator approaches. It is said to be altruistic because by making a noise the sentinel calls the predator's attention to itself. However, it is not clear where the reciprocity comes in—i.e. how it expects the non-sentinel individuals to repay its kindness. It is not really clear that it is being altruistic at all. It seems likely that if an eagle is seen approaching, by the time it reaches the spot the noisy sentinel will be one among a hundred pairs of fluttering wings, with only a small chance of being singled out. The selflessness of raising the alarm has been routinely admired but I have never read an account of anyone who has actually observed such a whistle-blower paying the supreme penalty.

The auxiliary argument is that it is at the very least sacrificing to sentinel duty valuable time that would otherwise be spent eating. It may be so, but the sentinel is usually a high-status individual which has priority access to the available food source and has presumably availed itself of that privilege. It may well be using its sentry-duty for digesting. Perhaps the alarm-raisers are not the bravest and most unselfish of the flock, but the ones with the fullest crops. It is at least a point that needs verifying.

3. Tick removal. This example figured in Dawkins's *The Selfish Gene*. It described a species of birds plagued by ticks; each one could remove them from all parts of its own body except the top of its head, so they performed that service for one another. It sounds reciprocal but not particularly altruistic. Altruism is biologically defined as performing a service for another which lessens, however fractionally, the inclusive fitness of the performer. There is no reason why eating a tick should lessen a bird's inclusive fitness any more than eating a caterpillar. The other point about this example was that it was hypothetical. If there had been such a species, Dawkins was suggesting, that is how it would behave.

His reasoning was so sound that although no such bird has been found, a species of impala has been identified which is indeed plagued by ticks and they are unable to reach the ones at the backs of their own necks, so they perform the service for one another. They do it simultaneously, so there is no time lapse and no risk of defection. The word for that is not reciprocal. It is mutual.

4. Care of the young. Helping a female to care for her young is usually done by her mate or other offspring, and is most easily accounted for as parental instinct plus kin selection. However in the case of the dwarf mongoose, babysitting is often performed by an immigrant female seeking to join the pack. It is suggested that this is Reciprocal Altruism; her own offspring may later be reciprocally cared for by those who gratefully remember her services. But they are more likely to be cared for by later immigrants who have no favour to remember with gratitude. It seems likely that the instinct to care for the young (related or unrelated) has been strongly coded for in the genes of this species, and that females who have no young of their own are prone to exercise it indiscriminately, rather than reciprocally.

5. Sparing a defeated enemy. It is sometimes argued that when a rival challenging a dominant male has been induced to concede defeat, and is allowed to crawl away and lick his wounds, this is an instance of reciprocity; the victor is sparing the life of his enemy so that at some future date the clemency will be returned. But the victor has gained his point, and pursuing it any further could incur some collateral damage even from a weakened foe. Only in chimpanzees and humans are such victims sometimes hounded to death.

6. There remains the vampire bat. These bats live by sucking the blood of other animals, and they need frequent sustenance. They can starve to death within 48 hours and around eight per cent of them on any given night may return to the cave without having found a victim. The luckier ones have often drunk more than they need and may disburse the surplus to a hungry bat in the vicinity. Their own offspring naturally have first claim on this service. Relatives may be favoured (kin selection) and acquaintanceship also seems to affect the outcome since propinquity within the cave no doubt influences the choice. After a hard night's hunting a bat returning to its roost is unwilling to seek far and wide among the hundreds in the cave for the most deserving recipient. It is more likely to disgorge to the nearest supplicant so that it can fold its wings and go back to sleep.

Careful research, taking pains to rule out the possibility of kin selection, has proved that Bat A is more likely to be fed by

Bat B if it has previously fed Bat B. However, it is not easy to distinguish this possible Tit for Tat element from the other known factors such as propinquity and acquaintanceship. I imagine if someone had saved my life by disgorging half of his last meal into my gullet, I would be inclined to count him thereafter among my acquaintances. However, if this is not reciprocity, it must be admitted as a border-line case.

Which raises another problem. If a gene for reciprocity was really part of our mammalian inheritance, it is very strange that it seems to have vanished, or failed to be manifested, in every species except two unrelated and ill-assorted ones: *Homo sapiens* and the vampire bat. Would it be possible for supporters of RA to work out what the two species have in common? That might give us the hint of a pattern and would help us to predict where we might find a third example.

As far as I can see, the concept of an instinct for Reciprocal Altruism makes no predictions, reveals no patterns, and explains nothing that cannot be explained more simply without it. All that is called for by way of explanation is the ancient, ubiquitous, hard-wired response of repeating actions that result in pleasing consequences and avoiding those that result in unpleasant ones. In the case of humans it is not necessary to stipulate that an altruistic action should be rewarded by repayment in kind. It may be adequately repaid with warmth, with smiles, with gratitude.

This explanation is a million miles from "blank-slatism." Humans are a social species and—other things being equal—their psychological well-being is heavily dependent on being accepted and approved by others of their kind. That dependence is certainly genetically transmitted, and is displayed widely throughout many social species. Babies in the first weeks of life can already distinguish a smiling face from a frowning one, even if depicted on cardboard in diagrammatic form—and they clearly show their preference for the smiles.

The reciprocity theory itself could not stand up for a moment without implying an original unprompted propensity to behave benevolently. If that did not exist, no one would ever have cause to show gratitude. If no one ever showed gratitude, RA could not exist. The theory that it is evolved behaviour is specifically predicated on the assumption that a genetic response of gratitude had already evolved—even though we hear no talk about the Grateful Gene.

The simplest explanation of altruism is a genetic instruction to the organism to desire good relations with others of its kind, plus a lifelong process of learning which kinds of behaviour will lead to that result in its own particular circumstances. To some scientists there is one major drawback to that formula. It allots too large a role to Nurture. The particular set of circumstances confronting any human being is subject to so many variables that it is next to impossible to do mathematics with it.

Yet Kin Selection had arisen out of mathematics, and was justified by mathematics. It seemed blindingly obvious that to put the coping stone on Hamilton's work, as Trivers set out to do, it would be necessary to get even more mathematical. We had to envisage a little mathematically-minded gene instructing its organism to start silently ticking off favours received and favours repaid, adding and subtracting, to determine what its next move should be.

Hence the excursion into games theory. There was never any pretence that the subject species was meant to be other than

Homo sapiens. But the characters we meet in Game Theory are not people. They are a special brand of humans who are never allowed to think or feel or learn from experience. Each homunculus taking part in the interactions has to be hard-wired to practise the same strategy at all times and in all circumstances. He must respond to all offences as if they were of equal gravity. If John's algorithm tells him to punish Chris when Chris has offended twice, then he must do so, regardless of whether Chris's offence was to knock his glass of beer over, or burn his house down. His response will also be quite unaffected by whether Chris is a nubile young blonde or a smelly old drunk or a six year-old child, and regardless of whether he himself is suffering the mother of all hangovers or has just won the pools and is feeling great. If the human race consisted of such individual organisms, the game would be able to teach us a lot about ourselves.

The victorious end to the PD marathon is particularly hard to map onto human experience. What does it mean to say that Tit for Tat works if the game is repeated, say, a hundred times? If the hundred repetitions are performed by a single player, then any human being who persisted in a strategy which had already proved disastrous more than ninety times would have to be seriously stupid, and have messed up his life beyond hope of recovery. If the implication is that playing the game a hundred times in succession is analogous to a hundred generations of humans, it still doesn't work. The second game would not be played by the same contestants. In the top left-hand corner we would have found the offspring of the miscegenation between John "Sucker" and Mary "Always Defect." And no such hybrid entity would ever be allowed to participate in a game of Prisoner's Dilemma.

In the face of such unanimous acceptance of the PD evidence by so many brilliant minds, I would hardly dare to make such statements, if it implied that everyone else was unaware of such mismatches between the game and the living world. That is not the situation. Everyone is aware of them. Here is Dan Dennett talking about Robert Axelrod: ". . .as he himself points out, the rule's provable virtues assume conditions that are only intermittently—and controversially—realized." What is not clear to me is how they can skate so lightly over that fact as if it didn't matter.

Perhaps I will be told: "The players cannot allow such niceties to enter into consideration because this is just a game, for Pete's sake!" My objections are seen as irrelevant, like interrupting a game of bridge to enquire whether the Knave of Hearts has been cleared of the tart-stealing charge, or removing a black bishop from a chessboard on grounds of simony. I can see the force of that. On the other hand, you do not normally emerge at the end of a bridge or chess tournament and write a paper revealing what the game has taught you about the psychological proclivities of kings and queens and clergymen.

The relevance of Game Theory to the evolution of behaviour depends on a long string of unverified assumptions. It assumes Trivers was right in postulating that reciprocity behaviour is inherited, not acquired. It further assumes that the different approaches to strategy—suspicious and spiteful or trusting and forgiving—are also inherited, not acquired. If these assumptions are not fulfilled, then there is no way of spanning the gulf between parlour games about Personified Strategies in "Darwinian" competition with one another, and the real world of flesh and blood.

I have read numerous accounts of the triumphant end of the game theory tournament, but in none of them has anybody condescended to add a sentence in plain English beginning: "What this proved about Trivers's theory was. . ." As far as I can see, the answer is "Nothing."

Consequently I was left feeling a bit like the child in Southey's poem about the Battle of Blenheim:

> "And what good came of it at last?"
> Quoth little Peterkin.
> "Why, that I cannot tell," said he.
> "But 'twas a famous victory."

Summary *There is no evidence that a human being's relations with other people are governed by an inherited mental module designed to calculate the chances of any unselfish action reaping a direct or indirect reward. "Do as you would be done by" is far likelier to be a learned response.*

"...the middle of the road is not the safest place to stand."

CHAPTER 8

THE TROUBLES

There is politics aplenty in Sociobiology, *and we who are its critics did not put it there.*

— J. Alper

In America, in the early 1970s, political feelings were running high. It was only a few years since the assassination of Martin Luther King and a lot of people were fired with the determination that his dream of racial equality should become a reality. Others were involved in a separate but equally passionate fight for Women's Liberation, and the two sets of campaigners had one thing in common. Both believed that the inequalities they were protesting about were social constructs, artificially imposed on one section of society by another section. So they both felt antagonism towards anyone who seemed to be saying that the inequalities were in some way preordained by biology or Darwin or nature. They suspected that these ideas were being used to imply that racial and sexual inequality was justifiable and, in any case, could never be eradicated.

In the 1970s, there were a number of scientists at Harvard who harboured that suspicion, including Stephen Jay Gould, Professor in the Museum of Comparative Zoology and the Professor of Biology, Richard Lewontin. Also at Harvard was Professor E. O. Wilson, working very hard on what he originally designed to be an advanced textbook for students, a general work of reference. He seemed only marginally aware of any ripples of social discontent. His book was published in 1975 under the title

Sociobiology: The New Synthesis. The subject was social behaviour in animals and it covered a very wide canvas. He himself had previously specialised in studies of invertebrates, so he was grateful to any of his Harvard colleagues who might offer guidance in the areas where he had no personal expertise, such as the social behaviour of *Homo sapiens*, as featured in the final chapter of the book. In the Preface, he recorded his gratitude to them and to one above all: "I am especially grateful to Robert L. Trivers for reading most of the book and discussing it with me from the time of its conception."

If the book had been 30 pages shorter, it would probably never have led to the furore that caused it to be splashed all over the media, but the last chapter dealt with humankind and began: "Let us now consider man in the free spirit of natural history." Others before Wilson had viewed man in that free spirit and it had led them into trouble. Thinking about human organisms primarily in terms of their genetic endowment makes it very easy to conclude that black people and female people and poor people must have inherited lower IQs than prosperous white males; that some people are just born to be losers; and that the rational way to organise society is as a meritocracy with the cleverest people on top.

These suggestions were unpopular. Publications by people like Arthur Jensen and Richard J. Herrnstein had been greeted with outrage, "wanted" posters, and public demonstrations in Wilson's own university. Wilson was not in fact travelling down that path, but he should not have been quite as surprised as he seemed to be by the events which followed. One of the themes of the last chapter of the book was that several existing academic disciplines were obsolescent and ought to be scrapped, and that sociology, for example, was in the natural history stage of its development, i.e., where biology had been before Darwin and Mendel, and was getting nowhere.

Other subjects due to be scrapped included ethics. "Scientists and humanists," he wrote, "should consider together the possibility that the time has come for ethics to be removed temporarily from the hands of the philosophers and biologized." To some of his colleagues, it may have seemed that he was trying to edge them out of their jobs, remove their departments from the prospectus and institute a new structure with biologists and geneticists in the driving seat.

Wilson earnestly believed that he was saying something new and important. He was still urging in 1982 that the "is/ought" distinction is unnecessary and should be eliminated as soon as possible. He realised that this proposition might be initially resisted or misunderstood, but he once commented that a certain amount of controversy helped to keep his adrenaline flowing. Possibly he quite looked forward to reading some crisp critiques of his ideas in future copies of professional journals. What he got was the sky falling on him.

The backlash provoked by *Sociobiology* began with a letter to the *New York Review of Books* denouncing Wilson's publication on political grounds. It recalled the events that had followed from the earlier advocacy of sociobiology in Germany, with references to sterilisations, eugenics, gas chambers, racism and genocide. It also dismissed Wilson's conclusions as bad science. It was signed by a list of people with impeccable academic qualifications, including his Harvard colleagues, Richard Lewontin and Stephen Jay Gould.

From that point on, opinions on the subject were increasingly polarised. Political groups who were already distrustful of the 'biology is destiny' attitude were driven into postures of extreme behaviourism. There was not quite 100% unanimity on the matter on the Left. Noam Chomsky, despite his unimpeachable left-wing credentials, declined to get involved. But in the atmosphere of the time, the protestors felt that any publication laying so much stress on genes as the determinants of human behaviour could have disastrous social repercussions.

A lot of the people who shouted protests and carried banners had not read the book and had a limited grasp of the principles involved. Wilson's denial of any political motive behind the book was as sincere as it was unavailing. Almost all people on both sides of this kind of controversy are convinced that their opponents are politically motivated while they themselves are merely seeking and defending the truth.

At the height of the controversy, the American Association for the Advancement of Science organised a two-day symposium on the subject. For one session, contributions had been invited both from Wilson and from Stephen Jay Gould, but before Wilson could begin to speak, chanting broke out in the audience. A group of dissenters mounted the platform and seized the microphone. Water was certainly involved. Some have reported that the contents of a glass were thrown over Wilson, but the incident grew more dramatic in the telling and it has definitively gone down in history that they poured a jug of water over his head. Gould had to use all his eloquence to restore order by condemning such tactics before the debate could proceed.

Lewontin planned a book-length response to *Sociobiology* which was published with S. Rose and L. Kamin in 1984 under the title of *Not in Our Genes*. But at the height of the polemics most of the commentators displayed knee-jerk reactions and shot from the hip. The politics of the Cold War got into everything; charges that the protestors were simply parroting Marxist dogma were countered by reminders of biological determinists who had joined the Nazi party and defended ethnic cleansing. Both sides trawled through history, public and private, in search of ammunition. Charges that "you used to be a social Darwinist" were countered by charges that "you used to be a Lysenko-ist." Every review was immediately scoured for the answer to the one burning question: "Whose side is it on?"

In the United Kingdom, the reactions were somewhat less extreme. A couple of top-line scientists with left-wing sympathies, while admitting to initial gut reactions against Wilson's ideas, resolved to keep politics and science apart. Peter Medawar stressed the point that IQ can never be separated into genetic and environmental components, since the same genes may be expressed differently in different environments. But at the same time, he deplored the conspiracy theories which ascribed malevolent intentions to Wilson and his supporters and was dismayed when they were vilified and shouted down. However, when feelings are running so high, the middle of the road is not the safest place to stand and Medawar was promptly attacked by combatants of both sides, including Wilson himself.

A few months after the appearance of *Sociobiology*, Richard Dawkins in England published his own book. It contained no chapter on *Homo sapiens*. But, like Wilson's book, it reflected Hamilton's new gene-based approach and made it, for the first time, accessible to the general reader. And the first edition had a Preface by Robert Trivers, which struck a more polemical note than anything in the actual text. That was enough to make Dawkins a target. He was surprised to detect a note of hostility in the questions that followed some of his lectures, and at one stage he was attacked by a series of letters in the *Guardian*. Showing commendable grace under fire, he remained silent until one letter levelled a charge that he found intolerable; it named him as an enemy of humanism. He allowed himself eight words of rebuttal, "I am Vice-President of the British Humanist Association. R. Dawkins."

It was not science's finest hour—on both sides far more heat than light was generated. It is easy to see in hindsight that the tactics on the Left were counterproductive, but from where they stood, it seemed that if people were induced to believe that human behaviour was hard-wired, the world would be a worse place to live in. People could lose hope; the injustices they had dreamed of mitigating would become set in stone. If I had been there at the time, I have no doubt that I would have been carrying a banner. In one way, the Left's instinctive response to genetic determinism was like the bishops' angry reaction to Darwin, though neither side would relish the comparison. "If this is true, let us at least hope it will not become generally known."

One unfortunate result was that the events left the door open for enemies of reason to advance the theory that science itself has nothing to do with truth; that scientific ideas simply float along on the surface of the tides of history; that, like any other myth, they are responses to primal urges wrapped up in a decent covering of self-delusion and rationalisation. A scientist has to have a core of optimism and a robust faith in his vocation to remain impervious to that siren song.

One man who had those qualities was Richard Feynman. "All sorts of people come up with ideas," he said, "and to a true scientist, it doesn't matter who they are and it doesn't matter what their motives are." In the end, when the idea has been examined and talked over and ways of testing it have been proposed and implemented, we will all end up a little wiser than we were before. And in the long run that is, more or less, what happened. Everybody cooled down and everybody learned something.

Stephen Jay Gould, without saying he had been wrong in any particular, admitted to tactical errors: "Our rhetoric was at fault." Richard Dawkins, also without saying he had been wrong in any particular, conceded that one much quoted image about lumbering robots had been a bit too melodramatic, "a rare purple passage." E. O. Wilson worked his way back into public favour by, among other things, his work on a project which was dear to his heart and incontestably on the side of the angels, the preservation of biodiversity.

In a newspaper interview in 2001, he said "I think the sociobiology controversy is essentially over" and in the same interview recalled, "I came from the old South, I was raised as a racist, I mean we all were." To my mind, when anyone says, "I was a racist" in the past tense, that wipes the slate clean. If I had been born white in the Southern States and given that upbringing, I would have been a racist. Anyone on the Left who cannot face that truth about himself comes dangerously close to claiming to have anti-racist genes in his DNA. I don't believe in those, any more than I believe in the tooth fairy.

The *Sociobiology* controversy seems, in retrospect, to have been in essence yet another replay of the nature-nurture debate.

Like the Lamark wars it receded into history and the political heat went out of it. What exactly was agreed on was left unspecified. Perhaps it was a reacceptance of the fact that the interdependence of heredity and environment is absolute: you can't have one without the other. Perhaps arguing over their relative importance in determining human behaviour is as futile as asking whether the oxygen or the hydrogen is more important in determining the nature of water.

Or perhaps it was merely a recognition that the polemics were doing everybody more harm than good. For whatever reason, one science journalist summed up the position by writing: "Unusually enough, this dispute seems to have been resolved to the satisfaction of both sides." But Ullica Segerstråle, who wrote by far the most complete and well balanced account of those events, was more cautious. "It may be a hasty conclusion," she wrote, "to say that the sociobiological controversy in a moral/political sense is over."

In recent years, a polemical note has once again been creeping into discussions of these topics. The question is whether the current disputes are merely aftershocks of that scientific earthquake, or whether seismic pressure is once again building up along the same fault line.

> **Summary** *Scientists frequently assert that science is purely about "is" and not about what "ought" to be. But occasionally the message that trickles through to the outside world sounds like "What you are hoping to do cannot be done, believe me." And since politics is the art of the possible, that is apt to be regarded as a political message.*

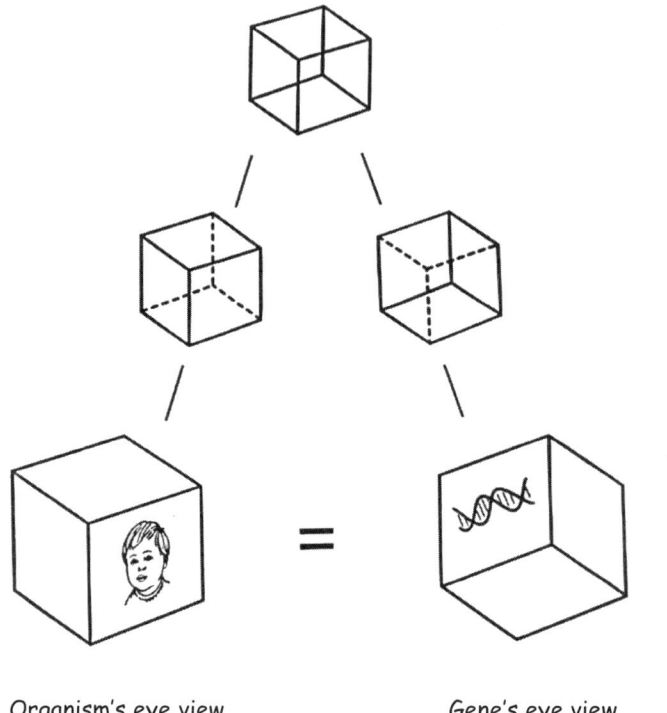

Organism's eye view   =   Gene's eye view

The point is that neither of the two perceptions is the correct or "true" one. They are equally correct.

Richard Dawkins

Chapter 9
Genes and Memes

Let us try to teach generosity and altruism, because we are born selfish.

— Richard Dawkins

Richard Dawkins's book *The Selfish Gene* had more impact on the general public than Wilson's *Sociobiology*. Far more people read it, and felt they understood it, and incorporated it into their view of the world. Dawkins presented the evolutionary process in clear bright colours. He got pleasure out of contemplating it, and like all first class educators enjoyed the experience of causing the same light to be switched on in other minds. He had given a course of lectures on Hamilton to students at Berkeley in 1968, and found it exhilarating to communicate the gene-centred perspective to undergraduates at a time when it was new enough to come as a revelation. But his book was aimed at a wider audience, and a different kind of teaching was called for: a style with fewer obscure abstract nouns, more striking phrases, more vivid analogies. Dawkins knew how to supply that—no man better.

To make his message accessible he dramatised it by personifying the influence of heredity on behaviour. Such personification is a common device and saves a lot of time. Take the goose that pulls an egg back into the nest. You can say explicitly, every time: "There is a tendency to retrieve eggs because the geese that practised this manoeuvre—even if originally it was a rudimentary or even accidental gesture—left more living offspring than the ones who didn't." That is long-winded and gets boring. Instead, for the sake of argument, you can say that the goose's genes "want" or "instruct" the goose to retrieve the egg.

Dawkins knew, and clearly said, that of course the genes are only dollops of DNA and have no conscious "desires" or "intentions." Furthermore there is not one gene to one piece of behaviour. Even in respect of physical characteristics, it is not that simple—for example it takes more than 72 chromosome sites to determine the texture of a mouse's pelt. But to simplify the picture still further, Dawkins singularised the genetic components concerned with behaviour and arrived at the term "*The Selfish Gene.*"

It was vivid and effective. But it made it harder for the untrained mind to hang onto the idea that this striking character—which played the title role in the book—was a figure of speech. For some people it fell into a groove in minds prepared to receive it by decades of science fiction spine-chillers, about people turned into zombies by aliens who took their minds over—and those entities always had desires and intentions, usually wicked ones.

Dawkins was unhappy when some people among the wider audience reacted adversely, as if they were allergic to his message. They wrote to him describing how they had spent sleepless nights after reading his book, how some of their students had been reduced to tears. "I am almost," he reported, "driven to the despair of which I am wrongly suspected." Why were they behaving in this way? Perhaps some of them were believers in the supernatural, and he had upset them in the same way that Darwin had upset their forefathers? As an outspoken atheist he would have had no difficulty in living with that. He is one of the very few public figures with the courage to protest, when television channels or reputable newspapers present items about astrology and satanic forces and the unquiet dead with an air of sagacious inquiry, instead of classing them where they belong, among fairy stories and superstitions.

Or perhaps the opposition to *The Selfish Gene* was political? Undoubtedly some of it was, at the time of the Troubles. From the internal evidence of his books, Dawkins himself appears to be apolitical to the point of innocence. He doesn't see where the trip-wires are and which turns of phrase will put people's backs up. But though that might account for the people who stood up and shouted, it would not account for the ones who lay down and wept.

Perhaps they were people so ignorant of science that they could not follow the reasoning behind the book? Some of them could certainly follow it, but still were not happy with it. Professor Randolph Nesse, who specialises in medical aspects of evolutionary biology, remembered in 1994: "The discovery that tendencies to altruism are shaped by benefits to genes is one of the most disturbing in the history of science. When I first grasped it, I slept badly for many nights, trying to find some alternative that did not so roughly challenge my sense of good and evil."

Perhaps they were just wimpish Pollyannas, believing that all the world is sweet and all stories will have happy endings? Bewailing any attempt to take their comfort blankets away from them? Well, you cannot halt the onward march of truth for people like that. This one seems to be the explanation that he settled for. And on that basis Dawkins stuck to his guns. "Such a very proper purging of saccharine false hopes, such laudable tough-mindedness in the debunking of cosmic sentimentality," he wrote, "must not be confused with a loss of personal hope."

He seems to have underestimated the power of his own authoritative eloquence. It persuaded many people who lacked his educational background that the story he was telling was the *whole* story. If it had been the whole story, we would be left to raise again the old Shavian lament about hogwash. No brave causes any more. No heroes, only hypocrites. No brighter future since the genes are in the driving seat, and "as things have been, things remain."

This is the reaction that Dawkins feels is a harsh misjudgement. He does not see the world as a barren place, but a fascinating one—because *his* heart, too, leaps up when he beholds a rainbow in the sky. In his book *Unweaving the Rainbow*, he puts up a strong case for saying that by understanding *how* and *why* we see that beautiful arch in the heavens, we will find even more in it to marvel at, not less. So how can anyone think that the scientific viewpoint is depressing? He seems to be saying, with Stevenson, that "The world is so full of a number of things, I'm sure we should all be as happy as kings." That power to wonder at the natural world is something that we are all born with, and some of the best scientists and some of the best artists keep it all their lives. David Attenborough, questioned once about his love of animals, responded that he was not over-fond of them: "I am merely astounded by them." But after childhood, the wonder of it all can begin to fade into the light of common day for most people, and even for some poets. When Wordsworth wrote "The rainbow comes and goes, And lovely is the rose" he followed it a few lines further along with the words "*And yet...*"

For most people, for good or ill, what makes the greatest difference to how they feel about life lies in their relationships with other people. Almost as much as they need food, they need a basic allowance of human contact—affection, gossip, laughter, someone to listen and sympathise. And Hell is other people too. They can be very alarming. They can hurt you, and you can hurt them without meaning to. It is a minefield, and when it goes badly wrong the rainbow cannot help you much. And it is this dimension of life which the gene's-eye view and the robot image may seem to drain the blood out of, and leave us in a world more alarming than ever. It is not our genes that communicate with one another. It is whole organisms, in their infinite variety, and it is unsettling when they are rendered transparent or seen as the servants of their selfish genes.

The scientist cannot afford to be influenced by the possible effects of his work, but it is unreasonable to be surprised when others are. To some readers, the book reinforced the "all-against-all" impression which had dismayed people in Darwin's day. As Lewontin pointed out, if we regard others as basically hostile to us and behave to them as if they were hostile, that perception can soon turn into reality. Despite Dawkins's assurance that we can decide to defy our genes, there was no guarantee that we would succeed. To some his message sounded like: "If other people seem hostile, the hostility is real, it goes right down to bedrock. If they seem kind, it may be just a cultural convention, a façade." Dawkins had not said that, but perhaps he took insufficient precautions against leaving that impression with people who failed to read the small print (or had only read the title page).

As it happens, I lost no sleep over *The Selfish Gene*. I read it with great interest and pleasure and no desire to pull the covers up over my head. Perhaps being educated in other disciplines takes the edge off the conviction that there must be just one right answer to everything. When the Jane Eyre story was retold from the point of view of the madwoman in the attic, or Hamlet rewritten from the point of view of a couple of courtiers, nobody felt moved to argue passionately in defence of the Jane's-eye view or the Prince's-eye view. New perspectives can be regarded as illuminating rather than threatening.

The Selfish Gene was a new way of looking at evolution. Samuel Butler performed a similar manoeuvre in a minor way, when he said that a hen is merely an egg's way of making another egg. It was witty and it was true. But it did nothing to affect the fact that an egg is merely a hen's way of making another hen. Both statements are equally valid. The process is cyclical.

Dawkins himself made it crystal clear that that was what he had been doing, by using the analogy of the Necker Cube. That is "a line drawing which the brain interprets as a three-dimensional cube. But there are two possible orientations of the perceived cube, and both are equally compatible with the two-dimensional image on the paper. The point is that neither of the two perceptions of the cube is the correct or 'true' one. They are equally correct." He was encouraging his readers to look at things from a different angle. Offering a second perspective on anything is an immensely creative thing to do. It is like being endowed with binocular vision for the first time. The world becomes suddenly three-dimensional, deeper and clearer and easier to understand.

Dawkins later extended the idea in a way that makes the individuals more transparent than ever. Human organisms—i.e. people—had once been credited with making changes in the environment, and thereby exerting influence over their own genes. For example, if you introduce agriculture and become a farmer, you have less need of speed for chasing game, and more need of a strong back for digging. The selective pressures on the composition of the gene pool will be altered, and we might conclude that the actions of people have altered them.

But Dawkins asked the deep question: Who or what is it which caused the people to make the changes in the environment? One answer could be the gene. It "reaches out through the body wall" and ensures that the environment will be altered in ways that will increase its chances of long-term survival. In this version, organisms are merely its vehicles and its tools. Dawkins was thinking about weaver-bird nests and beaver dams, but if you go far enough in that direction, then Beethoven and Shakespeare and Michelangelo and Einstein also become transparent. You can argue that it was in fact The Gene that invented agriculture, composed the symphonies, wrote the sonnets, sculpted the statue of David and told us that e equals mc squared. Clever gene.

In recent years, there is some danger that the power of binocular vision which Dawkins bestowed on us is being lost. That is not because his new perspective is losing its hold on our imagination. It is because Dawkins has moved closer to claiming that it is *the only right* way of looking at things. His tone has changed almost imperceptibly from the hypothetical to the prescriptive. In *The Selfish* Gene he assured us that the views from the two perspectives were "equally correct." In *The Extended Phenotype* he declares that the individual organism "should be" (not can or may be) "thought of as a vehicle for replicators," and that the Necker Cube analogy "may be too timid and unambitious." In the 1989 edition of *The Selfish Gene* he is no longer saying maybe.

He no longer believes that "you can flip from one to the other and it will still be the same neo-Darwinism." On the contrary, "The Necker Cube model is misleading because it suggests that the two ways of seeing are equally good." Anyone who flips the cube is looking at it "the wrong way up." There seems no particular justification for making this change, other than the effect of constant repetition on the minds of his readers and the way they responded to the message. After conferring the boon of opening a second eye for us, he now seems to be trying to stick a patch over the first one, leaving us monocular once again.

On balance it would probably be better to discard the cubic analogy altogether. Life processes are cyclical, not rectangular. The hen produces the egg produces the hen. The gene perpetuates the organism perpetuates the gene. (Genes, not people, can replicate. People, not genes, can copulate.) If we replaced the image of a cube with the image of a sphere it would be harder to accuse anyone of looking at it the wrong way up.

The tradition of air-brushing all mention of the human organism out of scientific prose is an old and honourable one. In writing up an experiment, it is taboo to say "I put the crystals into the beaker." The use of the scientific passive voice ("The crystals were placed. . .") is obligatory. I can wholly appreciate the purpose of this convention. But I become very uneasy when the active voice is arbitrarily restored and the subject of it is not a person but a gene. Genes in this scenario will "take whatever steps lie in their power" (no need to resort to the passive: "steps will be taken"). They will "ensure their survival" or may "merely change partners and march on." Polar bear genes "can safely predict" that the future environment of their unborn survival machine is going to be a cold one. Repeatedly we are warned that this language is not to be taken literally, but the cumulative effect of it is insidious. Genes are the doers. Organisms are the done-to.

As we move deeper into the personalisation process, "memes" are also put into the nominative. We are talking now about things like songs and slogans, ideas, catch-phrases, ways of making pots, religions, words, fashions in clothes and body language—anything at all which passes from one human mind to another. According to N. K. Humphrey they "should be regarded as living structures, not just metaphorically but technically." They are not, like us, mere containers or vehicles, so they can be said to *do* things. They can claim brilliant successes, they can enter into competition with one another, they can take steps to secure their own perpetuation. These units of speech and knowledge are, as it were, hovering in the air around us looking for a vulnerable brain to burrow into and replicate.

Richard Dawkins invented the word "meme." He has explained that the original didactic purpose of the meme was the negative one of cutting the selfish gene down to size. Humans are affected by culture as well as heredity and he wanted to make the point that the genes cannot account for everything we say and do. He could perhaps have simply said that. But it might have detracted from the beautifully consistent approach that had been sustained throughout the book, if The Organism had suddenly undeleted itself and stepped out of the shadows and claimed to be the subject of all the verbs. To write, "Humans learned to light fires. . .*Homo sapiens* began to fashion tools. . .People sang songs. . ." would have made them much too solid and opaque, as if they belonged in a different book altogether.

There had to be another way of expressing these things. He had a model—the gene—before his very eyes. If a gene's-eye view had proved tenable, why not a meme's-eye view? Then the tone and the style could be preserved. The *idea* of making a tool, the act of firelighting, the words of the song—these things could become the active agents and the subjects of the sentences. The organism could continue to be the done-to rather than the doer. The attentive reader should already have grasped that this mode of expression was only a rhetorical device. As one commentator expressed it, Dawkins never disguised the "as-if-ness" of his concepts.

But the memes escaped his control and ran loose in the world, with dire results. We are now told that memes can reshape a human brain in their own interests. One of the ways in which they seemed to restructure some people's brains was to render them incapable of recognising an "as-if-ness" if it stood up and bit them.

The memes, like Frankenstein, rebelled against their creator. They whispered in their victims' ears that Dawkins was only a faint-hearted believer, and induced them to turn on him and charge him with pusillanimity. "I am occasionally accused," he reported, "of having backtracked on memes, pulled in my horns, had second thoughts. The truth is that my first thoughts were more modest than some memeticists. . .might have wished." But it was too late. Humans, already designated as the tools and servants of their genes, now began to be depicted as also the slaves of their memes. The study of human evolution was being transformed into a masochists' paradise. Genes-and-memes became coupled together, like Hamilton-and-Trivers, as if they had to be accepted or rejected as part of a package deal.

The hallmark of this double-whammy approach is the way it comes down hard on any fairy tale about a human being making a decision about anything. Susan Blackmore, for example, is fairly scathing about our talent for self-delusion: "We do not say to ourselves 'It is *as if* I have intentions, beliefs, and desires but 'I really do.'" Fair enough. The idea that we are all walking around in a state resembling post-hypnotic suggestion has been around a long time, and we are accustomed to it.

But it seems perverse that her friends the memes are subject to no such deconstruction. She doesn't write: "It is *as if* there were discrete units of culture definable as memes and *as if* they had desires." They are nowhere portrayed as zombies. They are full to the brim with intentionality and determination. "If a meme can get itself copied, it will. . .We may expect more and more people to become infected with memes that drive them to spend their lives propagating those memes. That is what memes do. . .The books, telephones, and fax machines were created by the memes for their own replication."

What is the evidence for suggesting that we are being *invaded* by a meme, rather than simply repeating or originating it? Oddly enough, it is the kind of evidence that scientists normally shun like the plague—namely, the subjective evidence from "how it feels." We are reminded of how it feels to be irritated by a catchy tune that we cannot get out of our heads even if we want to: we are its hapless victims. Some combinations of sounds, especially musical ones, do score deeper traces in the short-term memory than others. No doubt there is some physiological basis for this, as for the fact that after staring at a bright light the image remains for a time on the retina so that we continue to see the outline of a light-bulb even with our eyes closed. But no-one suggests that the light-bulb is hell bent on parasitising the optic nerve.

Within a short space of time, the meme for "memes" briefly infected a surprising number of minds. The philosopher Daniel Dennett is one who was fascinated by it. He approached it by walking round it and looking at it from all possible angles—in a word, with circumspection, finding it "distinctly unsettling, even appalling" and admitting that "I am not initially attracted by the idea of my brain as a sort of dungheap in which the larvae of other people's ideas renew themselves." Later he identified it as a handy term for a salient cultural item, and on that basis "interesting." He was of course fully aware of the "as-if-ness" of memes, but he reminded himself that many of the best ideas "must grow out of something *quasi*-, something *as if*..." So what might grow out of this one? Perhaps a whole new academic field of cultural cladistics?

In Dennett's hands, however, human thralldom to the meme is never as absolute as it became in the hands of more extreme converts. At the very least, he observed, something must have happened to transform our species from unwitting hosts of the mind-invaders into "witting hosts"—and there is a world of difference between a witting host and a parasitised lump of grey matter. He makes the important point that if our role was simply to make and pass on copies of the memes that are competing for our attention, we have turned out to be singularly ill adapted to perform it. Our "brains seem to be designed to transform, invent, interpolate, censor, and generally mess up" the material before passing it on. Just look at those lovely verbs, all referring to activities of human beings, and you will be reassured that the organism has not been totally erased from Dennett's image of the

Necker Cube. He domesticated the meme, and turned it from something appalling into something that can be reconciled with what he calls his "cock-eyed American optimism."

Personifying The Gene was an effective thought-experiment which freed us from a too narrowly organism-centred way of looking at the world. That was legitimate. Genes do exist. They have location; they have dimensions. Laymen may waffle about them, but the scientist can precisely determine the point at which the waffle loses all contact with reality. He cannot do that with memes because they are purely abstract. To me, "fighting" and "loving" are verbs, describing ways in which people behave. To the meme-mongers they are also entities with the "as-if-ness" of a Will to Survive. To the Romans, they were a god and a goddess called Mars and Venus.

That is what worries me. I am afraid that if we get too much into the habit of attributing intentionality to abstractions, it may be a way of re-admitting the supernatural into our minds through the back door.

> **Summary** The Selfish Gene *was a brilliant rhetorical device for making Hamilton's concept accessible to the layman, but the gene's-eye view is no more valid than the view through the eye of the organism. As a tool for thinking about culture, the meme concept has not proved productive up to the present time.*

...the bloodlust of the early hominid males.

Chapter 10
The Pleistocene Inheritance

It's happened. We have finally figured out where we came from, why we're here and who we are.

— L. Betzig

By the 1990s, it was becoming clear that the gene's-eye view, if not the only game in town, was putting up a strong bid for that title. One triumphalist claim was that Wilson and his fellow researchers have essentially won the debate against "Gould and his loose confederation of academic allies."

Wilson had proposed that this new approach should permeate the thinking of academics in studies other than biology. One field of research which presented itself as a suitable case for treatment was psychology. It was an extremely soft science. The mind of man is an excellent tool for thinking about the world around us, but it is not so well designed for thinking about itself, and there was no single established methodology for tackling the problem. There was an anatomical approach with wonderful procedures like electro-encephalography and brain scanning, throwing light on the physiological hardware. There was a pathological approach attaching diagnostic labels to the ways in which the mind malfunctions. There was a therapeutic approach, trying out various cures and recording whether or not they worked, and there was a biographical approach, seeking connections between mental disorders and childhood traumas.

It was difficult to make it a harder subject by introducing more mathematics, because thoughts and feelings are almost impossible to sub-divide into standard units which can be counted. The maths that could most readily be used was of the actuarial type, as used by insurance companies. "If you do this or this, you are statistically more likely to die young"—and that involved counting real people, whole organisms rather than their genes.

In tackling problems where they cannot use arithmetic, scientists grope around for analogies, the "as-if" approach. For example, a flood of light had been thrown on the vascular system, when it was pointed out that the heart behaves as if it were a pump. If the heart is like a pump, what is the brain like? As they sat at their computers pondering about this, the answers came to them through their fingertips. In some ways, they decided, the brain behaves as if it was a computer. To a non-scientist it might seem that it was the computer which resembled the brain, rather than vice versa. There was no chicken and egg problem here. It was obvious which came first; human brains had devised computers to imitate some of their own mental processes. It was a bit like making the discovery that a man has two arms and in size and position, they bear a really remarkable resemblance to the sleeves in his coat.

However, the "as-if-ness" of this analogy was soon forgotten. The statement was modified to 'the brain is very much like a computer' and nowadays, nobody turns a hair when this appears in the terse form of 'the human brain is a computer.' Actually some people do turn a hair. There are people who look at the brain and compare it, not with human artifacts, but with other organs of the body. If you tell them how the brain functions like a computer, they will tell you the ways in which the brain functions like a gland. They too may reduce their view to the shorthand declaration, 'the brain is a gland.' The two statements are equally valid in their own terms, but the people who make them start from such different premises that they cannot communicate with one another.

It was time to find a new approach; a Darwinian approach which would introduce some order and discipline into the subject, using the golden key of Hamilton-and-Trivers. This was launched in 1992, in a book entitled *The Adapted Mind: Evolutionary Psychology and the Evolution of Culture*, edited and introduced by John Tooby and Leda Cosmides. They describe evolutionary psychology, EP, as a way of thinking about psychology. In this view, the mind is a set of information-processing machines that were designed by natural selection to solve adaptive problems faced by our hunter-gatherer ancestors.

Criticisms of that proposition have been dismissed as vapid pap and described as the ravings of incoherent environmentalists. So let us begin as coherently as possible by listing some of the basic assumptions on which the theory is based, and which no Darwinist in his right mind is likely to challenge. Evolutionary psychology is based on the fact that, in the animal kingdom, behaviour patterns can be inherited, as is the birds' ability to build a nest; that these behaviours are in most cases clearly adaptive and that if they seem not to be adaptive now, they may well have been adaptive at some earlier stage of evolution. That must be true of humans as of any other animal, so some aspects of our behaviour, as well as our physiology, may be best understood as relics of a prehistoric stage of our evolution.

The question arises: Which prehistoric stage? Our species has lived through many such eras and we bear the hallmarks of most of them. There must have been an era, for example, which determined that nowadays we sleep at night and are active during the day, although there is reason to believe that the earliest primates were small and nocturnal. Our babies respond to sudden startlement by a clutching movement and a hand grip strong enough to hold their own weight, which is probably inherited from a time when their ancestors lived in the trees and their ability to get a grip on their mother's fur was a matter of life or death. Again, Darwin pointed out that a human sneer has the effect of uncovering a canine tooth and may well be a relic of a far distant time when the tooth which would have been revealed was long and sharp and intimidating.

But the proponents of EP needed to home in on one particular period which they could regard as the EEA, the Era of Evolutionary Adaptedness. Their chosen EEA for the human race is the Pleistocene, the period after our ancestral line diverged from that of the apes.

The Pleistocene is thought of as a period when early humans were hunter-gatherers. Already they differed significantly from the other apes. They had certainly become bipedal and perhaps also naked, for reasons which are still hotly debated and will not be debated here. The advantage of choosing the Pleistocene is that it lasted a good long time, from about 2.5 million years ago to about 10,000 years ago, quite long enough for some behavioural reactions to have become fixed. The disadvantage is that this environment did not remain static throughout that period, which included four or five Ice Ages and inter-glacials, causing drastic climate fluctuations that may have necessitated migrating and changing their means of subsistence from time to time. Game hunting may have alternated with scavenging or fishing or nomadic pastoralism more than once before the advent of settled agriculture. But here again, for clarity and simplicity, it is easiest to create a single image and stick to it.

Tooby and Cosmides propose as "the most reasonable default assumption" that the interesting complex functional design features of the human mind evolved in the time of the hunter-gatherers. This assumption was not entirely novel, as there had been a 'man-the-hunter' paradigm in the 1960s, the brainchild of Raymond Dart, who was given to claiming that we owe our essential humanity to the bloodlust of the early hominid males and the weapons with which they would gouge out an eye or rip up a belly. Some people chose to regard EP as simply a new and less gory version of that. Indeed, E. O. Wilson felt that there was nothing particularly new in evolutionary psychology at all; that it was a recycling of his own Sociobiology, and its sponsors should have acknowledged it as such, rather than giving it a new name and trying perhaps to distance themselves from the hostility that the word 'sociobiology' still sometimes evoked.

However, it was quickly hailed as a new discipline. It made the cover of *Time* magazine as sociobiology had done. It also seemed to do more than any development since the 1970s to raise the temperature of scientific debate.

It is not quite as easy as it looks to decide whether an example of a fixed human behaviour pattern is a Pleistocene heritage; strictly speaking any such claim must be judged by three criteria. (1) That it must be found in all human cultures; otherwise it may not be fixed at all. (2) That it is not found in other primates, otherwise it might have become fixed long before the Pleistocene. (3) That it is not the kind of behaviour which would clearly be in our individual interest to practise in the modern world. Otherwise it may simply be the exercise of common sense or a learned reaction.

Evolutionary psychologists are generally very scrupulous about the first and second of these criteria. They recognise that some features, like incest avoidance and infanticide, should be handled with care, since they are common features in many other species, including strictly vegetarian ones. That indicates that their origins go back very much further than the proposed EEA and that hunting and gathering may have nothing to do with the case. The third criterion is the one that is most often violated. Philip Kitcher in 1985 had deplored the way that sociobiologists leaped to embrace an evolutionary scenario without considering the credentials of obvious alternatives, and evolutionary psychologists were not immune from that tendency.

A star example of this is the EP approach to mating behaviour. Statistics show that most men in our society, when looking for a marriage partner, tend to choose women younger than themselves and that women are equally predisposed to look for an older man. It can be readily shown that, while this is not true of other apes, it can be found in all human cultures and it is not hard to think up good reasons why a Pleistocene hunter would have been well advised to choose a young bride, since she would have more fertile years ahead in which to bear him many healthy offspring, and for the Pleistocene female, it would be adaptive to accept an older man. He would be a more experienced hunter and would have had time to achieve high standing among his peers.

Does not all this amount to a virtual proof that such mating practises arose in the Pleistocene and that is why we behave in the same way today? Not in the least. The same choices would be made by anyone contemplating any agreement with a second person, which would involve long term interactions and a shared lifestyle—for example, anyone looking for employment as a personal secretary or entering domestic service. Other things being equal, they would prefer to work for someone with experience and status and wealth and these things tend to accrue to older men. Equally the prospective employer would prefer someone eager and malleable and energetic and not too set in their ways, and if possible, personally presentable. If the object is matrimony, the "easy on the eye" factor is of major importance and younger women are prettier.

Ah but, we are cautioned by some evolutionary psychologists, you have not thought this thing through. They have tested their thesis experimentally by showing pictures of women to men and found a very powerful tendency for them to declare that the young ones are prettier. But their searching question is, *What makes them imagine that the younger ones are prettier*? They have clearly inherited the tendency to think so, because it was adaptive in their Pleistocene ancestors to be subject to that illusion. Another EP volume even attributes it to "a beauty detection mechanism specifically designed for rape."

I would dearly like to believe this theory: that in some underlying cosmic sense, I am just as beautiful as a woman half my age and only this biological blip renders so many male humans blind to the fact. But in terms of consilience, we have now strayed from biology into aesthetics and this gives me the confidence to call it nonsense. Show the young men in your experiment photographs of roses; they will prefer the newly opened ones to those which have stood around in a vase for ten days and begun to wilt. That is not because this choice would have increased the inclusive fitness of their hunting forebears. And the aesthetic preference for symmetry (also dubiously canvassed as the side effect of a desire for a healthy breeding partner) is manifested in innumerable ways that have no conceivable connection with choosing a mate.

The final argument against the theory that women are hard-wired to be sexually attracted to older men is the fact that, when they are sufficiently rich to have the power of free choice of a partner for sex, they do not hire older men. They hire toy boys.

Another aspect of EP which lacks clarity is the supposed mechanism by which the Pleistocene inheritance is handed down to us. The key word here is module. Module is the term used to describe a packet of information stored in the brain. These modules are not fictitious; they exist. No one has ever seen one, but that is no argument against their existence. In Mendel's day, no one had ever seen a gene, but he was able to infer that something of that kind had to exist.

The brain does have a habit of storing behavioural information into just such packaged units: otherwise we would find it very hard to make our way around the world. The first time that you ride a bicycle, you have to make a conscious effort to control the various muscles of your body, so as to keep upright, learning moment by moment which movements are most likely to achieve that end. When you have done it often enough, your brain wraps that information up in a bike-riding module and the minute you get into the saddle, the module is activated and you go through the appropriate sequence of movements without thinking about them. It has become, as we say, second nature.

Walking is second nature to us, but even though we have been doing it for six or seven million years, it has not yet become first nature. The toddler is not born with such a module; like the cyclist, it creates it by trial and error. A spider on the other hand is born with a behavioural module for spinning a web and has no need to create it. The two processes are entirely different. Yet whenever EP identifies a behavioural module it treats it as inherited rather than acquired, without first examining and eliminating the other possibility.

No one criticised EP writers for the complexity of their hypothesised modules, because most people had already swallowed the Trivers module. His RA theory postulated a hereditary pack of instructions encapsulated in human DNA, containing an order to the unborn, "When you get out of here, start keeping track of people. Get to know one from another and use the enclosed putative 'exchange organ' to keep account of how often they have repaid any favours you do them, and how often they have defaulted, and then treat them accordingly."

That is quite a package, and once it has been assimilated, no one is going to quibble about a simple command module, like "marry a woman younger than yourself." Some of the inferred modules are so complicated that you might be inclined to wonder how they get translated from the module into the behaviour. But that question is outside the remit of evolutionary psychologists. It is the province of the people who work at the messier end of human biology, the anatomical end. They will tell you that conveying messages about choosing a mate is not at all like saying: "Double-click on copulation and press Enter." It involves the gonads and the amygdala, synapses and enzymes and hormones, elations and agonies and frenzies. Some of these intermediary factors are not even in place at birth but are installed later, individually bespoke, tailored to fit the immediate circumstances of the developing organism.

Tooby and Cosmides are looking at the question through a different lens. "In this view," they wrote, "the mind is a set of information-processing machines." That is all it is: it is a view. It is an "as-if." It says in effect, "If the brain were no more than a set of information-processing machines, the following conclusions could be drawn." As with the gene's-eye view, they themselves, as well as their readers, are liable to forget the "as-if-ness" in the excitement of playing with the idea and finding new patterns in the data. People who sign up to their premise tend to adopt an evangelistic tone of voice; they have seen the light, they are spreading the message and those who fail to embrace it must be either thick or prejudiced. It's an easy frame of mind to fall into; I have been there myself.

In the case of EP, opinions are hardening on both sides. Non-scientists are being drawn into the argument because they feel their professional status is being unwarrantably attacked. The rival factions are no longer depicted as Right and Left, but given new and changing labels. The next two chapters illustrate this by examining two specific social problems, which evolutionary psychologists feel they can help to solve.

Summary *Evolutionary psychologists represent their new discipline as the next logical step along the road of the gene-centred view of human nature, and sometimes imply that no other view is an option for people who call themselves Darwinists.*

...the same difficulties beset biological fathers.

Chapter 11

Cinderella

Having a step-parent is the most powerful risk factor for severe child maltreatment yet discovered.

— **Martin Daly and Margo Wilson**

Once EP had been promoted as a new way of thinking, hopes were raised that it might be able to throw light on the ultimate causes of problematic types of human behaviour. The methodology used in these exercises is to collect a mass of statistics about the way humans actually behave, to provide some accounts of animal behaviour to give it a Darwinian perspective, and if possible add some reference to ancestral life on the savannah. An early and controversial example of this was a small booklet entitled *The Truth about Cinderella: A Darwinian View of Parental Love*.

It is a study of abusive step-parenting, which it describes as a "non-adaptive or maladaptive by-product" of the way the Pleistocene moulded our evolved psyche. The Darwinian element in this particular study is somewhat perfunctory. We are reminded that a male lion, on taking over a pride, kills any existing cubs to ensure that the females will cease lactating and become sexually receptive again as soon as possible. Other animals, including male langurs, behave in the same way.

But this has little or no relevance to the human race. Such behaviour is unknown among any of our nearest relatives the apes, and is also extremely rare among monkeys. (The langur is a notable exception). It is not easy to see what the lifestyles of hominids and langurs ever had in common, to indicate why persecuting step-children would have been more adaptive for them than for the rest of the anthropoids.

However, the statistics about human behaviour in the modern world are given in great detail. Daly and Wilson provide evidence that, compared to children brought up by their own parents, step-children are far less likely to love and be loved; that their presence tends to destabilise marriages; that they leave home sooner and are in much greater danger of being neglected and/or maltreated and/or murdered. In one study of child murder in Canada, a co-residing step-parent was shown to be approximately seventy times more likely to kill a child under the age of two, than was a co-residing parent. Other instances rate the increased risk factor far higher. The most extreme figure reads: "The odds ratio for this particular kind of lethal assault by step-fathers versus genetical fathers was approximately one hundred and fifty." These are startling figures.

In a pattern of events that has by now become familiar, the book encountered instant opposition and its conclusions were challenged. The authors reported: "We have been met with outrage and denial when we have reported them. The findings are routinely labelled controversial in media reports, and are sometimes indignantly dismissed as incredible."

Where were the outrage and denial coming from? The arguments in *Cinderella* cannot be called politically incorrect; they are not depicting step-parents as an innately delinquent minority. Quite the contrary, they are talking about you and me and themselves, and all of us, implying that if we allow ourselves to be cast in the role of step-parents, there is a high risk that our naturally sweet dispositions will be soured and we might become violent.

Daly and Wilson introduced a new note into the sociobiological debate by identifying the nay-sayers as being, on this occasion, not politically motivated but *professionally* motivated. They felt they were being attacked by "the practitioners," i.e. the social workers who were dealing with these problems on the ground, and might have been expected to be grateful for any guidance that was being offered to them.

The authors were scandalised when they found that, in a report of the AMA giving a list of abuse/risk factors which family physicians should screen for, step-parenthood was not even mentioned. They interpreted this as a disposition to hush up the true facts, and responded quite understandably by raising the tone of their claims to make them harder to ignore. That may have been one reason for the attention-grabbing quote on the cover of the Cinderella book, about step-parents constituting "the most powerful risk factor for severe child maltreatment yet discovered."

The social workers are inclined to see things from the point of the view of the organism rather than of the gene. The units they deal with on a daily basis are flesh and blood individuals, usually living in a sub-optimal environment which has left its mark on them. The kind of figures that would be helpful to them in their work would answer questions like: "Exactly what percentage of step-parents is abusive?"

The Truth about Cinderella is stiff with statistics, but this one is unaccountably missing. There is not even a guess at it. We are given to understand that offenders are in the minority, since "the relationship usually works out reasonably well"—but how small is the minority? Forty per cent? Twenty per cent? Three per cent?

Secondly, step-parental performance is measured in the book against the performance of the optimal alternative—a home with two genetic parents in a stable ongoing relationship. But the professional trouble-shooters are more liable to be called in when that optimum is not on the cards, because the genetic father has absconded or died, or cannot be identified, or is married to someone else. In statistical terms therefore they need to know whether the danger to a child from a step-parent is greater or less than the danger of a single mother sinking into a vicious spiral of debt, loneliness, alcohol and drugs, so that her child ends up in an institution. But in the book these comparisons are not made.

Thirdly, the practitioners have a quite different approach to history from the nativists (as they are currently called). Practitioners tend to think of history in terms of the last few hundred years rather than the last few million. From that perspective, the statement that the greatest danger to a child comes from a step-parent is simply untrue. Throughout most of recorded history, when a baby was killed, the murderer was most likely to be its mother.

Medieval cities often had a piece of waste ground outside the city wall, where unwanted newborns could be secretly taken and exposed to die before their mothers had time to bond with them. In later periods, young country girls might deliver in a ditch and cover the baby with leaves, or young urban skivvies in the lavatory of a railway terminal, where the infant would be wrapped in newspaper and dumped in a bin. Infanticide, like the stealing of other people's children, came to be regarded as one of the few crimes more commonly committed by women than by men.

Was it their genes that told them to kill? According to Hamilton, the kin selection machine should still have been working away at the old stand, saying "I order you to love that baby and devote your life to it." But the order was not obeyed.

There are two ways of explaining it. The believers in EP would simply put another tuck in the brain module. The gene's instruction must have been conditional: "Love your baby, but only if you can afford to rear it." That's possible. You can easily find examples of birds and mammals who neglect the runt of a litter to increase the others' chances of survival.

But the environmentalists favour a simpler answer—that the Hamiltonian maternal instinct was overborne by irresistible social pressures. The girls committed their crime because being known to have borne a bastard would mean ignominy, ostracism—even by their own parents—the end of any hope of a decent marriage or decent employment, and long years of grinding poverty for mother and child. The practitioners would argue that if the prevalence of that kind of infanticide has decreased, it is because of a change in social attitudes toward illegitimacy.

The same questions might be asked about the step-parents as about those desperate mothers. Why do some abuse while others do not? Under what circumstances are they most likely to offend? Perhaps it is significant that this is one other section of the booklet where the statistics dry up.

Daly and Wilson did note in passing that low-income families were over-represented in the A.H.A. data-set of child abusers. But we are given no hint of how much the figures are affected by income level. Could poverty be a cause of child abuse? Would a man's temper be more likely to snap if the family was living in a bed-sit rather than in a home where the children were cared for by a nanny in a sound-proof nursery?

Daly and Wilson dismiss the suggestion. "This initially plausible hypothesis was rejected on the grounds that the distribution of family incomes in step-parent homes in the US was virtually identical to that in two-genetic-parent homes." That statement is very carefully phrased. It does not deny that poverty makes poor parents more likely to attack step-children than rich ones are. It only stresses that that is just as true of natural parents as of step-parents, and therefore it is irrelevant to the analysis they are conducting. They reasoned that any association between abuse and poverty was independent of (was "orthogonal to") the case they were making. That's a good word, orthogonal. It means your thinking is at right angles to my thinking, and never the twain shall meet.

And yet there is so much that both sides agree on. The most perceptive sentence in *The Truth about Cinderella* is this: "Step-parents are primarily replacement mates and only secondarily replacement parents." It is true, and there's the rub. The mating system in humans is unlike that in any other primate. Most primates are not monogamous. In those that are, like the gibbon, the male does not undertake an open-ended duty to invest his resources in raising the young and the female is not required to be an ever-present helpmate to him, rendering personal services and psychological support. In humans, as in no other species, mate and offspring are, to some extent, in competition for these attentions. It is a rivalry that is particularly likely to threaten the harmony between children and step-parents.

Doesn't that prove that the Cinderella syndrome is coded for in the genes? Not necessarily, because the same difficulties beset biological fathers returning home from service overseas, who may at first be regarded by the children as interlopers. The other argument against its being "in the blood" is that while step-parents are more likely to be abusive than biological parents, the parents with the best record of all are the adoptive ones, where nobody shares anybody's genes.

I don't know of any Darwinist who would deny that the predisposition to love one's own biological children is very powerful, underwritten by biology and culture, and enables most families with two biological parents to make a tolerably good job of a very onerous responsibility. A step-parent's desire to do a similarly good job is based on goodwill, affection for the biological parent, and any affection for the children that might develop over time. In most instances, these feelings prove adequate to the task, but if the going gets tough, that kind of motivation is more likely to crack under the strain. A man with a pain in his psyche, believing that the world is treating him badly, may react by trying to pass on the blame and the pain to anyone around him who is not likely to hit back. It might be his wife or the cat or it might be the step-children.

The Old Left's response would have been: "Yes, we are all agreed on that, so the obvious remedy is to try to make the commitment less onerous. How about better family allowances? How about more nursery schools? How about a less unequal society?" The questions sound quaintly historical at the present time, but some answers will need to be arrived at soon because currently, the old-style nuclear family seems in danger of disintegrating. That is an added reason for the dismayed reaction to the *Cinderella* book.

Within the last 50 years in the West, the chances of today's babies' genetic parents staying together until death do them part have been cut by almost half. That has involved the most rapid and profound change in the basic expectations underlying human relationships for many thousands of years. The reasons are complex, probably in the broadest sense economic, but they certainly do not include massive mutations in human DNA or the human psyche. There is no consensus yet as to whether this trend could be reversed and even less about whether it ought to be. But the *Cinderella* thesis appears to imply that when rates of remarriage rise, the child murder and child abuse figures should rise equally steeply. If that is true the outlook is grim indeed. Does the book offer any solution?

It has a section entitled, "Can We Help?" But the answer has to be no, not a lot. It does not, for example, propose that all single parents should postpone taking a new partner until their children have grown up. That would be the only sure way to reduce the incidence of step-parenthood, but they can see it would not be a popular suggestion. The chief proposal made is that it might be helpful if "the step-parent's ambivalent and sometimes aggrieved feelings were acknowledged as normal and if the genetic parent were encouraged to express appreciation for step-parental investment, rather than to demand it as one's due."

Excellent advice as far as it goes; everyone likes to be appreciated and appreciation cannot be too warmly expressed. But it is remarkable that when the book moves from diagnosis to remedy, the problem to be tackled has suddenly shrunk from murderous hatred to ambivalent and sometimes aggrieved feelings.

The other problem is that you cannot send a message to the biological parent, "You must remember that your partner cannot help feeling resentment and anger, it is perfectly normal," without also sending a message to the step-parent, "you can't help what you are doing, it is perfectly normal, it goes back to the Pleistocene." That is a basic problem for all those who adopt the EP stance. They insist that they do not confuse what "is" with what "ought" to be, but in practise, they do tend to identify what "is," or what they conceive to be the case, with "what we can't do very much about, let's face it."

None of this implies that the practitioners are any wiser or nicer than the evolutionary psychologists. They have been prone to swallow some very dubious propositions in the past, and undoubtedly some of them do wish they could shout down Daly and Wilson. It is because they feel the book's lessons are unnecessary (everyone in the business is acutely aware that step-parenting can be a risk factor, even if they don't talk about it), and worse, that they can be counterproductive. On balance, step-parents are more often part of the solution than part of the problem, and demonising them, or being perceived as demonising them, could be a grave tactical error.

So far, despite the rising rate of reshuffled marriages and partnerships, fears that the child murder rate would rise to match it are mercifully not being fulfilled. In some American schools, in sectors of society where divorce rates are highest, more than 50% of a class of children may be living with only one genetic parent. Once their situation has become as common as that, they seldom report feeling disadvantaged by it. The kind of pressure that drives people to despair and violence often builds up as a result of feeling trapped in an unbearable situation. Nowadays, fewer men and women feel quite as trapped as once they did; more options are open to them. In earlier years, when marriages finally broke up, it was a last resort after years of hatred and bitter reproaches and unforgivable insults.

There are major disadvantages to having marriages that break apart more easily, but there is one advantage. They are likely to leave the partners less psychologically damaged, more resilient, sometimes on tolerably friendly terms. Step-fathers, at one time, were embittered by feeling downgraded and scorned by their peers. ("By God, I wouldn't stand for having to pay out all that money to bring up some other man's kid.") But as their experience becomes more common, they no longer have that extra cross to bear ("You, too? Join the club!"). The human race has passed through worse crises than this and survived.

The vastly increased awareness that children are often abused does not necessarily mean a vastly increased incidence of abuse. It was always there, but remained a dirty secret. Revealing it is the first step towards doing something about it. At any given moment, hundreds of thousands of individual human minds in different countries are addressing aspects of the changed situation, either on a personal or on a social level; thinking about play-groups, flexible working hours, legislation, women's refuges, car pools, subsidised childcare, counselling, networking.

If we are born with any hereditary factor that can contribute to the outcome of these efforts, it would have to be some putative gene for adaptability, the badge of all our tribe.

> **Summary** *Being a parent is stressful. There are many reasons why being a substitute parent is even more stressful. It is possible to clothe these facts in the EP language of inherited brain modules, but it adds nothing to our understanding of them.*

...to instil the habit of obedience.

Chapter 12
Rape

If young women really understood the evolved nature of male sexuality, they surely would be in a better position to avoid rape.

— R. Thornhill and C. T. Palmer

Two years after the Cinderella booklet, a more ambitious publication appeared: *A Natural History of Rape*, by R. Thornhill and C. T. Palmer. It is another example of the EP genre, and illustrates even more clearly the assumptions on which it is based.

The book sets out to challenge a blank-slate theory that rape is a political act, exclusively about power, that it has *nothing to do with sex*, and it would never occur to anyone to rape unless they had been taught it was the thing to do. Stated in those absolutist terms, it refers to a short-lived idea propounded in the 1970s as a kind of propagandist gauntlet thrown down in the heat of battle by a small group of extremists.

Others have told Thornhill and Palmer the same thing, but they refuse to be fooled. They suspect that those of us on the Left who don't swear a religious allegiance to the blank-slate paradigm are lying in our teeth. We only "*appear* to have disagreed"—and the italics are theirs. They need to set up this fairly easy target to shoot at, because they are going to come close to the opposite kind of extremism—arguing that rape is all about sex and has nothing to do with power.

However, they seemed aware that if they set up to combat the Feminist Menace in the year 2000, they were going to look a bit dated. So they announced that "We will refer to it as 'the social science explanation.' " Daly and Wilson's "practitioners" are identified not simply as social scientists, but more impressively as "the social science establishment."

The authors predicted that hackles would be raised by their book and that prophecy was apparently fulfilled. I gather they have been attacked for being macho, which strikes me as untrue. They sound perfectly humane and civilised people. I accuse them of nothing worse than a monocular gene-centred view of evolution which skews their thinking quite as much as the beliefs of the feminists were previously skewed. I also charge either them or their publishers with offences under the Trades Descriptions Act.

They claimed, or let it be claimed on their behalf, that they were offering remedies. Rape, said the book cover, could cease to exist once Thornhill and Palmer had revealed its causes. They were taking up arms against "a horde of humanists" and facing facts that would enable us to "stop rape."

Their approach to the problem relies on collected reports and statistics about human behaviour—in practise chiefly contemporary Western human behaviour—plus assorted references to wildlife. They place themselves squarely in the EP tradition by insisting in bold type that **the environmental problems our early ancestors faced were quite specific.** But they tell us not a word about what those problems were, nor what particular aspect of them rendered rape adaptive.

They lay great stress on their Darwinian credentials, and accuse social scientists of failing to "Darwinise." But they do not cope too well with the basic Darwinian questions: "What other species have evolved this feature, and what do those species have in common with *Homo sapiens?*"

They leave the second half of the question unanswered, and their answer to the first half is disingenuous. They claim that "the widespread occurrence of rape across animal species is both consistent with evolutionary predictions and devastating to the social science interpretation." They report rape among insects, referring in detail to certain scorpion flies and an insect called the water strider. They give a list of references intended to suggest that rape is common among fishes and reptiles, and among birds, marine mammals, and non-human primates. I am not going to argue about the water strider or the fishes. The birds referred to are ducks, and we will come back to them. But I totally reject the implication of the widespread occurrence of rape among mammals.

Craig Palmer, researching the subject in 1989, found that instances of rape had been observed to occur in two non-human mammals—elephant seals and orang-utans. In both cases, the males are at least twice as big as the females, and the social system leaves a high proportion of subdominant males permanently deprived of access to willing females.

With the exception of these two species, rape is not a recognised feature of sexual behaviour in any mammal other than man. There are plentiful instances of *aggression* against females in a number of mammalian species. It can take the form of unprovoked violence (often against females who are not in estrus, or are too young to mate) in order to reinforce dominance and instill the habit of obedience. It can involve chasing to the point of exhaustion, or sequestering and hindering the approach of any other male, or bites inflicted on the female after her successful resistance, as punishment or in response to frustration.

Thornhill and Palmer become irritated when people describe these behaviours as "sexual harassment" and demand to know why they are avoiding the term "rape." The answer is very simple. It is because in non-human species this type of bullying hardly ever culminates in copulation—and the reason for this is blindingly obvious.

Most mammals don't rape because they can't. It is virtually impossible for a quadruped such as, for example, a wildebeest to copulate with a female wildebeest that will not hold still and co-operate. All she has to do is to keep walking or running. Superior strength and size are irrelevant. Even speed is not particularly relevant. Having caught up with her he would have to stand still and rear up, giving her the opportunity to move forward a few feet and once again render his project null and void. If all male animals could rape, courtship behaviour would never have evolved. It would never have been necessary. Courtship is an appeal for co-operation. (This observation also applies to birds. Most of them can fly away from an unwelcome suitor. A duck floating on water cannot. It takes too much time and effort to get airborne.)

The second obvious reason why most male mammals do not rape is because they don't need to. Why not? Because the males are sexually aroused by olfactory signals from a female who is in a state of sexual receptivity. Her readiness to mate produces secretions which activate his readiness to mate. This beautiful system ensures that if she is not in the mood, he is not in the mood either. Sex in most mammals is a seasonal or cyclic occupation occurring perhaps once a year at the time when the females are in estrus. The estrus cycle does not dominate human sexual behaviour, and thousands of pages of research and theorising have been devoted to trying to explain why not. It is a fundamentally Darwinian question and the rape book treats it as not worth mentioning.

In non-human mammals the female almost invariably has the power of choosing a mate. This means that males (peacocks, lyrebirds, mandrills, and many species of fishes and even molluscs) are always the sex that dons the bright colours and the decorations, while the drab females look them over and judge them by their visual impact. Except for a handful of species with total role reversal, humans are the one exception to this rule. In humans it is predominantly the males who assess a potential mate by visual cues, and the females who dress in bright colours and rely much more on non-visual criteria for assessing the desirability of a male. Why has this extraordinary swapping of roles taken place? In the rape book the fact is not referred to and the question is not addressed.

Other unique aspects of human sex are referred to but explained incorrectly. In discussing female orgasm, the authors report that "a woman's frequency of copulatory orgasm is significantly predicted by the nature of the source environment she is in, and thus by the opportunity for successful parental effort." The phrase "significantly predicted by the nature of the source environment" is opaque. It may refer to the Pleistocene environment which in this volume has not been described. Or it may be an indirect way of saying "culturally determined." Culture certainly has a powerful effect on it.

In Victorian times, when women lay back and thought of England, top European scientists in the field were adamant in asserting that the female orgasm was a myth. Millions of women lived and died without ever experiencing it or hearing about it, and were contentedly unaware that they might be missing out on anything. Something must have radically changed since the time when it was selected for, to render its operation such a hit-and-miss affair in our species. What could it have been? Does its hit-and-miss operation indicate that it is still incipiently evolving or obsolescent and on its way out? In the book the question is not addressed.

It makes just one suggestion: that the purpose of female orgasm was to promote pair-bonding, by increasing sexual appetite in the female. Nothing is less likely. Increased appetite might promote sexual activity, but it is certainly not known for promoting monogamy or fidelity, either in humans or other animals. Among primates, the species with the strongest pair bonds (the gibbons) are the ones with the lowest sex-drive.

One question that is explored at some length is why women dislike being raped. It is rather a strange question. It seems a bit like asking why a man dislikes being mugged. Men too are sometimes raped but there are no statistics about whether or how much they dislike it. Thornhill and Palmer have assumed that the adverse reaction is peculiar to human females, and they hypothesise that the mental and physical distress women suffer after being raped is *adaptive*. Natural selection has decreed that they should suffer in this way.

Why? The suggestion is that since rape does not aid their inclusive fitness—they do better by carefully selecting the fathers of their children—the psychological anguish that follows it serves a useful Darwinian purpose. It "functions to guide cognition, feelings, and behaviour towards solutions." In other words, it teaches them to mend their ways so as to avoid a repetition of the events. I am not sure what our Pleistocene ancestors could have done to avoid a repetition, and the authors do not specify what remedial measures they had in mind.

Why have Thornhill and Palmer left so much of the relevant biological data out of account? It is because they are steeped in the EP tradition which teaches them to think about behaviour almost exclusively in terms of computational modules, "physiological mechanisms in the nervous system that, at the present state of scientific knowledge, can only be inferred from patterns of behaviour." But once you convince yourself that they exist, you find less and less need to wonder whether any of those patterns might be conditioned, or consequent on other changes. You postulate that each one has been selected for, and handed down intact, and that this formula will explain everything.

On this basis the suggestion is that a physiological mechanism has evolved in human males causing them to commit rape. I am not at all clear why a special module is necessary. Most rapists display the same mind-set as that of a mugger: "I want something; you don't want me to have it; but I am stronger than you so I will take it."

There is one other remarkable feature of these modules that has attracted little comment. Most men are not rapists and most step-parents are not abusers. So perhaps there is a hereditary module which dictates to them: "Show forbearance and protectiveness towards women and children," and the offending minority happen to be deficient in this module. The possibility is never even glanced at. That is one of the depressing things about gene-based theorising. It is almost as if there were a hidden agenda: to make us believe that the anti-social elements in our behaviour are always those with the deepest roots. That assumption is entirely arbitrary.

In short, Thornhill and Palmer have certainly made an effort to Darwinise this problem, but a little Darwinisation is a dangerous thing. If we stand back a bit from the statistics of current practise and view *Homo* as an animal among all other animals, a number of alternative questions and answers come to mind.

Why do men rape when other animals do not? They are more likely to be driven to it, because of the loss of estrus and because their lust is no longer conditional on signals of receptivity in the female. Secondly, men rape because they can—because grasping hands and face-to-face mating make it possible. (The other occasional mammalian raper, the orang-utan, has similar powers of restraint and the observed instances of rape are usually ventro-ventral.) Why in humans is it the females and not the males who are assessed for their beauty? Because once males became able to override female reluctance, females effectively lost the power of choice. Why is female orgasm not as reliable an accompaniment of copulation as male orgasm? Because the female behavioural reward had evolved over 60 million years to be triggered by dorso-ventral sex, and the physical reconstruction that followed bipedalism made it less easy to evoke by either approach.

I do not believe that rape was ever a "facultative adaptation" in the human male—i.e. a better way of doing things than the methods employed by other apes. The implied advantages offered by *The Natural History of Rape* would have applied equally well to apes and chimpanzees. It may have been something more like a measure of crisis management, a reaction to something that had *gone wrong*.

The authors may argue that my reaction to the book is influenced by the fact of my being female. Very likely it is, but only to the same extent that theirs has been influenced by being male. When discussing rape, views on the matter may well be influenced by whether you can more easily imagine yourself as the doer or the done-to.

There is one facet of human behaviour which seems acutely relevant to rape victims but which is not featured in the *Natural History*. It is the tendency—particularly strong in males—to divide humanity into antagonistic subsets, tribe against tribe, rich against poor, black against white, one football team or religion or nation or economic doctrine against a different one. It can lead to great heroism and solidarity and self-sacrifice; it can lead to lynchings and pogroms; it can lead to World Wars.

The point the feminists were making in connecting rape with power rather than sex is that the relationship between men and women is liable to become contaminated by the agonic behaviour patterns of the "us-and-them" instinct. That strikes me as a valid point. Men and women are certainly born different, but many women became convinced that culture has also superimposed on that difference a learned dimension of hostility: misogynism. That belief had political implications. If the attitudes to women—not individually but collectively—were at least partly cultural, those attitudes could to some extent be changed. Events have borne out that belief. The attitudes have appreciably changed.

When Thornhill and Palmer talk about rape they are not thinking about the same phenomenon that their critics are talking about. Their book concentrates on the lighter end of the rape spectrum, dwelling on the plight of hapless youths who have misread the signals or been falsely accused. The darker end is barely touched on.

The people who registered outrage at the book concentrated instead on the darker side. They thought about rapes accompanied by acts of humiliation and sadism and degradation; they thought of bands of youths setting out in the evening to commit gang-rape in exactly the same frame of mind as they set out for a spot of gay-bashing or racial harassment. They thought about dead children, and the fact that if a woman is found murdered in a public place, the first question asked by the police is whether there are traces of semen. Rape is sometimes about sex, and sometimes about power, and sometimes about a mixture of the two. *The Natural History of Rape* concentrates exclusively on the first kind.

Thornhill and Palmer try hard to make their approach less one-sided by introducing at the beginning and end of the book a token woman, who is either anonymous or hypothetical, and it doesn't matter which. She has recently been raped. It was, predictably, a rape on the mild side. A boyfriend got carried away, and afterwards begged to be forgiven. But she still feels bad about it. The authors cogently argue that it will not make her feel any better if some feminist comes along and says something like: "It was an outrage. He ought to be locked up or castrated. Men are all the same, all horrible, and in their hearts they all hate us." That might serve as an outlet for anger, but it would not be a recipe for happiness, and it would not be true. Men are not all the same, any more than women are.

If this troubled woman picks up this book it will promise to offer her advice about how to avoid being raped. If she reads it and is able to decode it, she may detect that phrases like the "guidance of cognition, feelings, and behaviour towards solutions" is a coded way of saying "You should wear higher necklines and longer skirts and never go out alone in the dark." That is prudent advice and, as the authors point out, she might have done those things anyway, out of fear. But she will also hear undertones of censure. ("You were really asking for it, weren't you? If anyone's behaviour needs modifying it is yours.")

She may recall that at one time all women wore skirts long enough to shield men against any inflaming glimpses even of their *ankles*, but that did not eliminate rape. She might reflect that if it is so very hard for predatory males to control their urges, she will only be saving herself at the expense of some other girl with a shorter skirt. It might occur to her that if her clothes are so prudent that her appearance will not attract the attention of any bad man, there is a chance that it will not attract the attention of any good one either, and that might not make her any happier in the end.

She might suspect that if this book was "A Natural History of Murder" or "A Natural History of Embezzlement" it would lay less stress on the faults of the victim. It would not point out that aggression and greed come naturally; they were selected for back in the Pleistocene. It would not say "If you feel badly about having your spouse murdered, your pain may be adaptive. It has *evolved*, to teach you to take better care of the next person you marry. And if you resent having your money conned out of you, you should reflect that you were really asking for it, putting temptation in the way of weak-willed racketeers by selfishly investing it where you thought it would attract a higher rate of interest, instead of prudently burying it in the garden."

The book will explain to her that male sexual appetites can be very strong and hard for them to control. She may think she knew that already, but she reads on because it has been intimated that this book is an eye-opening analysis which will lead to new strategies for dealing with rape and perhaps even eliminating it.

After all, the spokesmen for the "social science establishment" have signally failed to crack the problem. They resort to their customary orthogonal kind of analysis, pointing out that the incidence of rape is hundreds of times higher in deprived inner city areas than in prosperous suburbs, and the lower the income, the higher the number of rapes. They reason that one remedy might be to reduce inequality and end social exclusion. Failing that, they are limited to suggesting measures like improved sex education and appropriate legal penalties.

Finally she reaches the very last page of the very last chapter, and the long awaited section entitled "How can rape be prevented?"—the equivalent of Daly and Wilson's "Can we help?" Again the answer is no, not really.

The suggested remedies are that men and women should be taught about male and female sexuality; and men should be informed about the penalties for rape. Changes in the law should be based on scientific knowledge, but voters must decide for themselves what changes would be suitable. What's new in any of that? It is just another way of saying "improved sex education and appropriate legal penalties"—an unexceptionable programme which has always had the enthusiastic backing of everyone in the social sciences.

Finally the book reminds us that charges of rape may be false, because women sometimes lie. I think we can safely assume that people have already taken that on board. In Britain when a man is charged with rape, his chances of acquittal are in the region of 94.7%. Perhaps we need a reminder that women also sometimes tell the truth, and it is even possible that they tell the truth more than 5.3% of the time. If the effect of books like this is to convert 94.7% into an even higher figure, it will hardly be worth keeping the offence of rape on the statute book at all.

> **Summary** *In the great majority of mammals, rape is unnecessary and impracticable and does not occur. It is one of a whole range of ways in which human sexual interactions are anomalous and/or unique. A savannah environment in the Pleistocene fails to account for any of them.*

...they will shout at the television.

Chapter 13
The Origin of Empathy

Although male and female researchers do science the same way, they may be attracted to different problems.

— Sarah Blaffer Hrdy

It is time to return to the question of altruism and confront the challenge: If Game Theory isn't the key to the mystery, then what is?

It was a controversial problem for much of the last century. Where did loving behaviour come from, and when, and why? Konrad Lorenz thought it was a side-effect of aggression: "Loving arose in many cases from intra-specific aggression by way of a ritualisation of a redirected attack or threatening." Freud treated it as a by-product of copulation, suggesting that a baby's love for its mother was simply a rehearsal for a future sex-life. Trivers thought it originally evolved as a gamble on the hope of a pay-off, and Ghiselin contended that it was all a con, a socially-imposed obligation to pretend that any of us gives a damn for anyone's welfare except our own.

The odd thing is that Darwin himself and many of his followers saw no mystery at all. People are capable of being cruel and of being kind. Neither Love nor Hate appears to be calibrated through a computational module, since either may cause people to run amok until reason totters on its throne, and the organism behaves in ways that cannot possibly serve its well-being. But in the long run these aberrations are less counter-productive than being incapable of either aggression or affection.

The capacity to fight may be essential to self-defence and competition, and the capacity to cherish may be essential for raising the young. In mammals a division of labour has arisen so that the biological infrastructure for aggression is stronger in males and that for protection is stronger in females, but neither is restricted to one sex.

This way of thinking seemed to cover all the facts. It explained why humans generally believe themselves to be more merciful than other animals. English is not the only language in which words like "humane" are used to denote kindness, and words like "bestial" and "beastly" to denote ferocity. It was predictable that the empathetic instincts would be most powerful in the species whose offspring were born particularly helpless, and took a uniquely long time to become self-supporting.

Powerful emotions are often projected onto targets for which they were not designed. Ethologists call this "displacement activity." A man humiliated by his boss may take it out on his wife. People behave aggressively towards inanimate objects; they will shout at the television, throw a book across the room, deface photographs. Darwin believed that tenderness and unselfishness could be similarly diffused: "Woman, owing to her maternal instinct, displays these qualities towards her infant in an eminent degree. Therefore it is likely that she would often extend them towards her fellow creatures."

It was too obvious to need enlarging on. Professor I. Eibl-Eibesfeldt, in his book *Love and Hate*, still felt that way in 1970 and encapsulated it into seven words. Love, he wrote, had "arisen with the advent of parental care."

So when and why was this assumption examined and discredited? That is the whole point: it never was. There was nothing in Hamilton's paper that would undermine it. Parental protectiveness is the very apotheosis of kin selection. It might seem a bit wasteful to produce so much that some of it washes off onto strangers, but Darwin himself famously commented on the wanton wastefulness of evolutionary processes where they serve the ends of reproduction.

Nobody ever wrote a paper protesting that the belief in such diffusion was contrary to the known facts. There was just one thing against it: it is open-ended and unquantifiable. The individual organism looks for no return when it expends energy on its offspring, and when the same biological software is activated by other stimuli the same thing can apply. But altruistic behaviour that does not count the cost is out of tune with the zeitgeist, and the most elaborate circuits are resorted to, by the nicest people, in the effort to coax it back in again.

For example most scientists today accept that to succeed in the world of Prisoner's Dilemma and all its variations, you must be out to win at the expense of other people. They further accept that they themselves evolved to live in such a world. Yet when they catch themselves out doing benevolent actions—feeding the ducks, or sending a cheque to Oxfam—that theory does not reflect the way they feel.

Robert Frank looked for a way around that. In the world of Game Theory, in order to put one over on people, you have to induce them to trust you. But other people are so smart that the best way of convincing them that you have good intentions is really to *have* good intentions. That way, the hypocrites pretending to be nice guys may be out-selected by the ones who really *are* nice guys. If you think too long about that one, it makes your head swim. Entering into the Game Theory world and emerging again by a side exit doesn't seem to make things any clearer. It seems simpler to face the fact that animals, including humans, live in a world where the preconditions underlying Game Theory are simply not fulfilled.

The human brain cannot divorce itself from its biological history. It is simply one part of a large complex entity. The blood that flows through it is pumped by the heart; it carries hormones secreted by the adrenal glands and the gonads. It may be enriched, depleted, clogged, overheated or poisoned by events beginning with the eyes and hands and lungs and their intermediary effects on the stomach and the liver and the corpuscles. The relationship of these things to the brain is not that of obedient servants. It is one of interdependence.

Returning then to the original question: "When and why did loving behaviour begin to evolve?" the first problem is finding the right word for it. Many people are ill at ease with the word "love" and would prefer something cooler. "Altruism" has been too exhaustively argued over, and had too many different provisos attached to it by different people. The word charity doesn't feel right either, not since the dire phrase "as cold as charity" entered the language. I propose to settle for "empathy." It means literally "feeling with" and describes the phenomenon of being happy because someone else is happy and sad because someone else is sad.

If Darwin was right, we have to admit that empathy is a fairly late arrival among the emotions. Scientists used to be taught that all animal behaviour, including that of *Homo*, was governed by four primary urges. They might sometimes come into conflict, and different priorities could be allotted to them at different times. However, they were regarded as an irreducible minimum for steering any animal through the great battle of life—irreducible, but sufficient. To fix them firmly in their minds, the lads in the labs encapsulated them in an alliterative list. They were known as the four F's—fleeing, feeding, fighting, and fucking—and mediated by the well-known emotions of hunger, fear, aggression, and lust.

But at some point in the history of evolution a new primary urge developed. It began many millions of years ago with the first emergence of the paternal instinct—and no, that is not a typographical error. It began with the vertebrates, and the first vertebrates appeared in the water, and in the water it was more often than not the male that took on the responsibility of caring for the offspring. Up to a point, that tradition persists among fish and amphibians. A male cichlid will hang around, keeping predators away from the eggs he has fertilised and fanning them with his fins to keep them supplied with oxygenated water. A male midwife toad holes up in some dank hiding place with his eggs on his back to keep them moist and safe.

If you ask "Why the males?" you will be given the stock cynical selfish-gene solution, i.e. that in fish and amphibians fertilisation is external, so the female's egg-laying role is the first to be completed. She can then cheerfully make her escape and leave her mate to finish covering the eggs with milt, and then be saddled with looking after them. In mammals, the male is the first to complete his contribution, so he is free to get away and in most cases he does.

It is a plausible theory and I used to believe it, until I thought about the Emperor penguin, that most heroic of fathers, and realised it is purely the interests of the young that decides these matters. The male fish stays because, of the two parents, he is the less exhausted by his part in procreation and the better able to defend his brood.

Even in air-breathing species, the onus of parenthood did not immediately devolve onto females. For example, in a number of birds of ancient pedigree, like the ostrich and the megapode, the male commandeers the job of looking after the eggs. In migrant birds today rearing the young is usually a joint effort, since the rearing season is short and it takes two to complete the job in the time available.

Females became specialists in the parental role with the emergence of mammals. They evolved a physical modification which enabled them to continue to nourish the young after birth from their own physical substance. That modification (L. *mamma*, a breast) has given its name to the whole order. The process of giving sustenance in this way puts heavy demands on females. It calls on them to consume many more calories per day and at the same time hampers their freedom to move around in search of food. From that point on, the four F's list continued to be irreducible, but ceased to be sufficient. If any mammalian species had tried to get by on the four primal urges, however many fights were won and however many females were impregnated, the inclusive fitness of every individual would have been zero. All the offspring would be dead within days, unless the list was augmented by a fifth F: the instinct for fostering.

At what point did this basic drive begin to spill over into relationships with non-kin? Examples of this do seem to occur in non-human animals. Frans de Waal, in his book *Good Natured*, gives numerous examples to illustrate this. And within the last year or two a couple of striking examples have been reported. There was one spider monkey which reacted to the plight of an orphaned howler monkey with such a surge of maternal hormones that she lactated and was able to rear it. Other examples get wider press coverage, like the lioness in a Kenyan reserve which adopted an oryx fawn and "protected" it, even from its own mother, until it died of starvation. And there was the little boy visiting a Chicago zoo who fell into the gorillas' enclosure, and the nearest female gathered him up tenderly in her arms and carried him to where she could hand him over safely to the keeper.

There have long been reports of dolphins saving people's lives by bearing them up to the surface. Perhaps they are cued to respond in that way to any warm-blooded, air-breathing creature in that size range thrashing around in the water in apparent distress. The chances of its being anything other than a dolphin would be vanishingly small. And that kind of aid is not operative only between mother and child. Midwife dolphins customarily stick around when a baby dolphin is being born to ensure that it gets its first lungful of air in good time.

All this evidence may be dismissed as verging on the anecdotal. The hard evidence of a connection between maternal instincts and generalised empathy comes from anatomy, and as it happens the students of evolutionary psychology do not pay much attention to precisely how the body works. They make a statistical connection between a perceived pattern of behaviour and an inferred DNA-based instruction to activate that pattern. The nitty-gritty details of how the message is conveyed are irrelevant to their calculations.

But there are other scientists who find those details fascinating. Sarah Blaffer Hrdy is one of them. In her classic study of the female of the species entitled *Mother Nature*, she declared a personal preference: "I am interested in the organism"—or, in scientific terms, in the phenotype. And she gave her reasons. "The important point here is that all anyone ever sees, touches, or directly experiences is phenotypes, never genes. . .Only phenotypes are directly exposed to natural selection. That is why, evolutionarily speaking, and especially for those like me who study behaviour, phenotypes are what matters."

Among the anatomical features of the organism which may throw light on our evolutionary history are the hormones. The concept of hormones is by now familiar to the general public, but the names most familiar to non-scientists tend to be those associated with emotions of aggression and lust and anger, such as adrenaline and testosterone. Until recently we have heard much less about oxytocin. Women are more likely to have heard of it than men. Some will have heard it bandied about over their heads in hospitals where they may be encouraged to give birth in the daytime and on weekdays, rather than at unsocial hours like 3 a.m. on Sunday. An injection of oxytocin can help to induce a swift labour at a convenient time; it is also naturally produced in massive amounts in a mother when her baby is born and it is believed that one of its functions is to promote bonding between mother and child.

So there is a prima facie case for seeing it as a key agent in producing parenting behaviour. One interesting thing about it is the increasing realisation that it is involved in many other benign and pleasurable social interactions. Its production is not limited to the occasion of childbirth, nor is it limited to one sex. The secretion of oxytocin is stimulated, for example, by being massaged, by having an orgasm, by breast-feeding, by taking part in a convivial dinner party. It is produced in social animals when they groom one another. Sarah Blaffer Hrdy describes it as the endocrine equivalent of candlelight, soft music and a glass of wine.

It is intrinsically related to pair bonding. Sue Carter predicted that oxytocin would be found to be more important to animals which form long-term pair bonds, than to those species in which sex is a one-off encounter. She was vindicated when many more brain receptors for oxytocin were found in monogamous prairie voles than in a closely related species which does not practise pair bonding. If I had to make a similar prediction, I would guess that in a tear-jerking film like *It's A Wonderful Life*, at the climactic moment when all the problems are solved and all the people love one another, oxytocin is coursing through the veins of the viewers as fast as the lachrymal fluid is flooding into their eyes.

All this would support Darwin's hunch that an emotion instituted for one purpose may extend into other relationships and permeate many aspects of social life. The other significant thing about oxytocin is that we can roughly date its evolutionary origin: it is apparently a mammalian speciality. Many millions of years ago, some of the sweat glands in the skin on the ventral surface of some vertebrate species began to secrete a substance more nutritious than sweat, and these evolved into milk glands. Vertebrates with milk glands secrete oxytocin: vertebrates without milk glands do not.

It is easy enough to see the Hamiltonian mechanism which would sanction quite a strong dose of altruism in a female mammal; and to appreciate that it would have to be powerful. It would need to be robust enough to operate in competition with other pre-existing emotions, like fear and greed, otherwise a mother would desert her brood as soon as danger threatened. She would eat them if she got hungry; growing cubs would get nothing to eat until she had had her fill and if they pestered her, she might swipe them too hard and leave them maimed. Her love of them has to be strong enough to compete in that formidable gallery of instinctive behaviours. It does not always prevail; sometimes a female mammal does desert her young; sometimes she does kill them and sometimes even eat them. Sometimes she has a choice between raising one healthy survivor and starving the runt, or trying to raise two when resources are scarce.

In some social species, like the aforementioned langurs, where incoming dominant males destroy all infants fathered by their predecessors, mother love is strong. She will register outrage, she will strive to hide with the baby and if cornered, will shriek and protest. But it is not quite strong enough to induce her to pick up the infant and run away to a solitary life in the jungle. She needs the tribe around her; she will stick around and with any luck, her next baby will have grown up before there is another takeover.

Even in such extreme cases though, love is a contender; it has the gene on its side. Whether we call it altruism or love or empathy, what we are talking about is a relationship between individuals, a social relationship. In mammals, the primal social relationship is between a mother and her offspring. The male-female conjunction is of immense biological importance, but in most cases it is a fleeting encounter, no more in need of bonding mechanisms than an encounter between predator and prey.

In species where ongoing male-female partnerships do develop, they borrow the chemical messengers and the behavioural repertoire from the more primal relationship. The mother-infant bond is an unbalanced one; one partner puts into it more than she, as an individual, gets out of it. The reciprocal calculator, that exchange organ on which the whole apparatus of RA theory hinges, is quite irrelevant in this connection.

If parental instincts were indeed the original trigger for the emergence of unselfish behaviour, we would predict it to be strongest in species where the new-born are helpless and slow developing and have a lot to learn before they can survive on their own. This prediction is fulfilled. The new-born of *Homo sapiens* has a higher concentration of these attributes than any other. It is not surprising that we consider ourselves, as a species, to have a unique capacity for being empathetic, for being "humane."

The keyword is capacity; it is like the capacity for speech. A baby cannot talk if it has never heard speech; it cannot love if it has never received or witnessed loving behaviour. But the capacity for empathy is there in large measure; it co-exists with the other four F's. It has to be strong enough to compete with them, just as they compete with one another. It does not need, any more than they do, any elaborate computational formula to explain it away.

That perspective on human nature makes no claim to be a new way of looking at things. It is an old way of looking at things which has fallen out of favour, not because it has been invalidated, but largely because the new way is more fun to play mind games with. If we return to the earlier perspective, we find alternative accounts of many phenomena nowadays explained by

RA and EP. For example, the new versions tell us that the unique human capacity to read other people's emotions by studying their expressions must have evolved because, in the endless poker game between our computational brains, we need to be able to detect when others are trying to cheat us. It is at least as likely that it evolved because mothers and babies need to understand one another's needs and emotions during the crucial months before they can use words to communicate.

Some readers may feel that this chapter, ostensibly about the human race, has viewed it predominantly from the distaff side. It contains a lot of feminine personal pronouns and dwells on some features which are more characteristic of females than of males. That should not invalidate it. If it does, then there are hundreds of books and papers steeped in testosterone and cramped within the confines of the macho four F's, which would have to be discounted on the same grounds—namely, that they are concentrating unduly on the evolution of one half of the human race.

Matt Ridley wrote a whole volume about the origin of virtue, in which women are treated almost exclusively as (a) where egg cells come from and (b) the scarce resource which the real human race specialises in competing for. He devotes one sentence to them, in a *jeu d'esprit* at the end of a chapter. He is describing what the species might have been like if it had lacked the tribal instinct: "It is an intriguing fantasy to imagine ourselves like that. Indeed female human beings are like that already."

Unlike Ridley, I find the opposite sex intriguing, an endless source of mystery and fascination. I believe they deserve more than a sentence. They deserve a chapter to themselves. If I occasionally distinguish them from "ourselves," you will know where I picked up the habit.

> **Summary** *Erving Goffman wrote of parental behaviour: "There is no appreciable quid pro quo. Balance lies elsewhere. What is received in one generation is given in the next. It should be added that this important unselfseeking possibility has been much neglected by students of society."*

"Isn't it lucky it wasn't on his face?"

CHAPTER 14

IT'S A BOY

Better fighters tend to have more babies. That's the simple, stupid, selfish logic of sexual selection.

— Richard Wrangham and Dale Peterson

For any writer who is being honest, the mysterious sex must simply mean "the one I don't belong to." There are some biological arguments for regarding males as the more mysterious, inasmuch as femaleness appears to be the basic blueprint for a species and maleness the exotic deviation; for example if a frog's egg is stimulated into developing by a needle rather than by a sperm, it may produce a healthy frog but never a male one. Being female is the fall-back position.

In a male mammal, we are told, what the Gene cherishes about him is his aggression, which increases his chances of passing on his DNA. That was why it was felt that the continued existence of nice guys needed an explanation: how could the nice-guy genes have escaped being selected out? It might help to remember that every specimen, male or female, has a beginning, a middle and an end, and try to tell the story of a male from the beginning.

We are accustomed to reading that maleness is a consequence of the presence of a y chromosome in the DNA: xx=female and xy=male. But that is not an invariable rule. In birds it is the males that have matching chromosomes. In sea turtles, the sex of offspring is not determined until after the egg has been laid:

those incubated at low temperature become males and those at higher temperature become females. In slipper limpets your sex depends on how old you are, and you will get a crack at both roles before you die. In humans as in most mammals the y chromosome predestines its owner to maleness, but the above examples may make it easier to understand why it is some time before the gamete (the fertilised egg) pays any attention to that instruction. It is busy with more primal tasks—transforming itself from a spherical ball of cells into the semblance of a mammal. As far as sex is concerned, it leaves its options open for quite a long time, even to the extent of creating two rudimentary sets of sex organs side by side, one male and the other female, so that when the time comes to concentrate on that particular matter it can simply erase one set, rather than constructing the other one from scratch.

At this time, as at every other time in its life, its fate is governed by a mixture of heredity and environment, and one thing we all have in common is that whether we like it or not, we all spent our formative first months in a 100% female environment. Ancient philosophers used to teach that a female's only contribution to the next generation was to provide a container for it. A little homunculus, they said, was already alive and complete in its father's sperm, and its mother had no more influence on what it would grow into than the shape of a flower pot has on a hyacinth bulb. We now know better than that. We have long known that the mother contributes 50% of her offspring's genetic make-up, but we are still in the process of learning how much its fate may also be influenced by the state of the organism that is acting as its temporary container.

It has long been known that if the mother is undernourished, the baby is more likely to be born prematurely and to be of low birth weight, and to suffer the consequences of these conditions. We have recently learned a few other facts—and learned them the hard way. For example, the experience of the thalidomide babies revealed that sometimes drugs which are harmless to the mother can cross the placental barrier with horrific effects on fetal development. Nowadays, too, the mother of a baby with spina bifida can bitterly blame it—not as in the past, on some sin for which God is punishing her—but on something she ate.

So we have a steadily increasing knowledge of how a fetus may be influenced by foreign substances that its mother eats or inhales or is infected by while she is pregnant. The effect of some of them can depend on the stage of development of the fetus, so that for example German measles virus may be disastrous in the first three months of gestation and comparatively harmless later on. What we are still deeply ignorant about is how that development may be affected by fluctuations within the normal range of the mother's endocrinal status—in other words by how she is feeling and what is happening to her and how she reacts to it.

There used to be a series of old wives' tales about the things that a pregnant female should avoid. If she happened to see a hare in a field, it was said that the baby would have a harelip. There was even Biblical authority for the belief that if you arrange to keep pregnant ewes in a stripy landscape, they will bear blotchy lambs. All these superstitious beliefs have been discarded, and the advice offered by doctors nowadays about how to treat expectant mothers is relatively simple. See that she gets enough good nourishing food; yes, she can carry on working and she can still have sex; just urge her to cut down on alcohol and cigarettes, and try to see that she doesn't get upset.

That last bit is obviously endorsed by the Gene, because it is already busy pumping her full of Nature's own-brand sedatives. Her output of estrogen and progesterone starts to increase from the moment the embryo is implanted, and one of the effects is supposed to be that they help to lull her into a bovine placidity for the duration. Quite often they work, at least up to a point. The only thing nobody tells you is what happens to the fetus if she *does* get upset.

It is an almost impossible problem to research. Scientists can compile tables correlating babies' birth weights and their general fitness with maternal eating habits, their patterns of smoking and drinking and sleeping and general family circumstances. That method has come up with valuable results such as the recommendation of more unsaturated fats in the diet of expectant mothers. But you cannot produce a convincing graph by going around the maternity ward asking everybody: "Did you get upset at any time in the last nine months, especially in the early part—say in August of last year? Exactly how upset, on a scale of one to ten. . .?"

And yet it would be interesting to know more about this, especially in the case of the gamete with the y chromosome. Unlike his sisters, he is hopefully destined to end up with a hormonal balance quite different from his mother's. Some of the various aspects of maleness which are going to be successively switched on in the course of his development to maturity will be responses to hormones that he himself is going to have to supply. In the very beginning, all fetuses are effectively neuter. Even later, when a male fetus has developed some little gonads of his own and they are going into production, it seems possible that the effect of their output might be enhanced or counteracted by a sudden surge in the amount of testosterone or estrogen flowing through the umbilical cord and into his veins because his mother is in a state about something.

The question is worth raising because any discussion relating to male behaviour patterns in humans is liable sooner or later to run into the politically charged topic of homosexuality. Is it genetically determined or is it purely the effect of upbringing and social conditioning? The incidence of it doesn't fit very comfortably into either of these hypotheses. And because it is a subject that many people feel passionately about one way or the other, almost every finding about it is hotly contested as soon as it is announced. Some of the differences in the attitudes of Right and Left were predictable. Many on the Right were likely to be repelled by homosexuality, reacting to it with a kind of xenophobia as an unacceptable deviation from the norm. Many on the Left were inclined to rally automatically to the defence of any minority that was being discriminated against, as gays certainly were.

The surprising aspect was that the biological preconceptions of Left and Right were suddenly reversed. It was now the Right which denied that genetics had anything to do with the case. A right-wing father confronted with the fact that his son was gay (and gays, like accidents, will happen in the best-regulated families) would often agonise over "Where did we go wrong?" and end by blaming it on his wife, recoiling from any implication that this thing might have been inherited from him. Left-wingers on the other hand, the traditional upholders of Nurture against Nature as the determining factor in human behaviour, equated sexual orientation with skin colour as something people were born with and which therefore should never be used as grounds for prejudice. They welcomed Dean Hamer's claim to have found a "gay gene." They helped to publicise the fact that homosexual behaviour had been found in other species, like bighorn sheep and bonobos and a wallaby and a plover. Some assumed that if this behaviour was innate, it must have been at some stage adaptive, and they embarked on the ambitious task of inventing scenarios to show how being gay in the Pleistocene might have helped people to bequeath their genes to future generations. A certain amount of truth and a certain amount of nonsense emanated from extremists at both ends of the spectrum.

It seems quite possible that same-sex orientation could be innate (determined before birth) without being hereditary (determined at conception). Certain aspects of sexual characteristics are switched on at specific points in our lives, like the growth of beards or breasts. And some of them are switched at certain points in our pre-natal development.

To take one example, it has recently been announced that one lesser-known example of a secondary sexual characteristic is the relative length of the first and third fingers on each hand. In men the forefinger is typically shorter than the ring finger and in females the reverse is true. In my own case, I signally fail to pass this gender test, as well as the one about choice of toys. Why should that be? I do know that my normally equable mother could on occasions spectacularly lose her temper. When she was fifteen and saw through the window a baker's delivery man kicking his horse, she rushed out and laid into the man with the business end of a sweeping brush until it was wrenched out of her hands by a bystander. I have speculated that later on, when she was pregnant, she may have witnessed somebody being beastly to a pussy-cat, and that could have released in her an overdose of testosterone that miscoded the length of my fingers and robbed me of my rightful pleasure in playing with dolls.

Something similar obviously happened to Jo March in *Little Women*. But the tomboy syndrome is common enough in little girls to pass without much comment. If Mrs. March had got churned up a couple of weeks earlier, or later, Jo might have been a lesbian and then the sequel *Good Wives* would have been a quite different kind of book. The same thing could of course operate in reverse. It is quite on the cards that early in 1854 Mrs. Wilde had a beautiful and moving experience, with the result that unborn Oscar's veins were temporarily awash with high-grade oxytocin at a critical stage in his development.

However, for sociologists looking for ways of persuading people to live more harmoniously with one another, gays are not the main problem. The main problem is posed by the straight males who comprise 90% of the sex we share this planet with. These people, though they constitute less than half of the total population, commit 90% of the murders, and are responsible for the great majority of all the crimes in the book, except infanticide and prostitution.

You might think it would be women who would be most likely to highlight the brutality and aggressiveness of males, but that is not true. There was indeed one wing of Women's Lib which held that men were without exception SCUM, but they could never out-vituperate that sector of male evolutionists who took their cue from Raymond Dart.

Dart waxed eloquent about the males of our species, describing "the blood-spattered, slaughter-gutted archives of human history from the earliest Egyptian and Sumerian records to the most recent atrocities of the second world war" and the "worldwide scalping, head hunting, body-mutilating and necrophiliac practises of humankind in proclaiming this common bloodlust differentiator, this predaceous habit, this mark of Cain. . ." It was supposed to be because they started eating meat, and it was written, like a tabloid newspaper's six-page spread on the latest sex scandal, with a thin veneer of disapproval and an undertone of glee. Books delivering the message were in great demand, giving the impression that libraries everywhere were echoing to the drumbeat of male readers drinking in this portrait of themselves and thumping their chests. That may be a misinterpretation. Perhaps they were merely contrasting it with their own behaviour, and reflecting: "She should think herself lucky. Considering my baleful heritage, I am really remarkably restrained."

The good news is that most of the males we encounter in our daily round are not up to their slaughter-gutted elbows in mutilation and necrophilia. Any of them may (any of us may) be capable of doing terrible things, as the history of the last century revealed. But those things are not the human norm. Something is clearly counteracting the simple stupid selfish logic of sexual selection. To use one of E. O. Wilson's favourite metaphors, something is keeping it "on a leash." It is widely assumed that the leash must be made of culture, artificially imposed by society on a seething substratum of savagery. But that assumption may not be justified.

Suppose we return to our fetal friend and try to watch his synapses being moulded by events. Since he has not yet emerged, you might think events cannot have taught him anything up to this point. A uterus is not a very eventful place. But when something does happen in there, he is already capable of learning from it by trial and error. Perhaps one day while he is squirming around making aimless movements with his limbs, his hand may come into contact with his face. He is already hard-wired to respond to this touch. Touch a newborn's cheek and it will turn its head towards that side and close its lips around any suitable sized object that it encounters, such as a nipple or its own thumb, and will suck it and find the sensation soothing. That chance sequence of events leaves a little memory track in the brain of the unborn like the line in the sand where a trickle of water draining from the land makes its way into the sea. If the sequence is repeated the same track will be followed and deepened.

And these things can happen before he is born. His brain forms an association between the originally random movement and the pleasure of sucking and he will learn to control that movement of his arm and repeat it purposefully. By the time he is born he will be either an addictive thumb-sucker and remain so for the next few years, or a non-thumb-sucker who will never acquire the habit. Once he is out in the world he is too busy responding to the nonstop stream of events and sensory perceptions that bombard him throughout his waking hours, forming connections and creating a brain that will never be quite like anyone else's, even if he is an identical twin.

Building a cerebral cortex that works is largely a matter of laying down connections and circuitry—synapses—and every brain is bespoke: it is assembled to order, to meet the demands made on it, like the electrical wiring in a building. Le Doux expressed it by saying: "People don't come pre-assembled, but are glued together by life."

So the physical make-up of an animal's brain reflects its

needs. A South American monkey with a prehensile tail needs a fair amount of cerebral wiring devoted to operating it, and the connections between an elephant's brain and its trunk will monopolise a hefty percentage of its cortex. An ape on the other hand has no trunk and nothing much in the way of a nose; the functioning of its brain will differ accordingly. Some years ago, a scientific paper published an imaginary picture of the human body with the different parts of it proportioned to the percentage of the cortex devoted to operating them.

The figure was so bizarre that it was reproduced in the popular press. It showed a caricature human with a very small torso (most of the processes of our internal organs are not consciously controlled and take up virtually no room in the cortex). But it had monstrous hands, and monstrous lips, reflecting the vital importance to human beings of manipulation and speech respectively. It also had a monstrous penis.

What struck me at the time was that the configuration of a woman's body as perceived by her cortex must surely be quite different. A little girl lacks a penis the way an ape lacks a tail. The anatomical equivalent of the male generative organ—her ovaries—are buried deep inside her and she is no more conscious of them than she is of her liver or her spleen.

The full significance of this is partly obscured by the androcentric habit of describing the vulva as "a sex organ," simply because males have the unshakable conviction that there ought to be an organ at that location in every human being. But there isn't. An organ is "a means of action or operation; a person or thing by which some particular purpose is carried out or some function is performed." The female equipment which will one day perform functions lies deep inside the body wall. The vulva doesn't do anything. It is an aperture, like the earhole, and if a little girl's attention is not drawn to it, the one will loom no larger in her consciousness than the other. That would seem to leave a sizeable amount of cortical space free to be devoted to other purposes. Nobody has enquired as to what they might be—more refined social skills perhaps, or slightly greater linguistic ability.

It must be as hard for men to imagine what it is like being without a penis as it is for us to understand what it's like being with it. The classic illustration of that was Freud's famous assumption that women are cursed with an inconsolable sense of loss because Mother Nature has deprived us of that priceless boon. But Freud was brought up at a time when children were kept in sexual ignorance, when babies were either delivered by the stork or discovered by their delighted mothers curled up under a gooseberry bush. He had no excuse for not realising that millions of young girls who were not blessed with brothers could progress quite a long way towards puberty without even knowing that such a thing as a penis existed, let alone mourning the lack of it.

A story used to be told of a little girl who was amazed when one day a visiting neighbour undressed her new baby boy to change his nappy. The child was overcome with pity, and whispered into her mother's ear "Isn't it lucky it wasn't on his face?" Every woman hearing that story can testify how much more credible that is than the words Freud would have put into her mouth: "Please, mummy, can I have one of those?" Of course her response was inappropriate. Properly evaluated, a phallus is a splendid invention, without which none of us would be here today. Thinking of it as an excrescence would be quite as silly as thinking of females as if they were amputees. If the little girl had had brothers, and had been accustomed from birth to seeing them with no clothes on, she would have considered it as natural and becoming as the long ears on a rabbit or the antlers on a deer. *Vive*, as the French have it, *la difference.*

Now the boy child has opened his eyes on the world, and his brain is growing very fast and changing with his every waking hour, and it is being shaped by, and to fit the needs of, his environment. His synapses owe zilch to the Pleistocene. They are a "now" thing. They are responding to the contingencies of today's world.

Imagine a baby born to parents of hunter-gatherers with a Stone Age economy—some of them still exist. Imagine he is handed over at birth to a pair of prosperous, well-educated parents to bring up as their own. His chances of succeeding in life will be no different on grounds of origin than if he had been born to an urban couple from say Delhi or Omsk or Detroit. Suppose he was one of a pair of identical twins; they might share some of those believe-it-or-not parallels, popular with science journalists and the reading public. Perhaps they would both have a horror of spiders, and both bite their nails. But one of them might be a motor mechanic or a newsagent or an architect, and the other would still be shooting at monkeys with a blowpipe. Their lifestyles would have carved out their respective talents and behaviour patterns, and by the time they were twenty there would be no prizes for guessing which was which.

So we cannot yet predict what kind of Peter or John or Thomas this little boy will grow into. We couldn't do it even if he was born with a complete print-out of his personal genome strapped to his ankle. He finds himself in a totally new situation.

In some ways he is more vulnerable now than he was before he was born. For one thing he is no longer cocooned against the dangers of being scratched or neglected or dropped on his head. Also, before birth, at the level where genetics is the only determinant of the outcome of conflicting interests, the fetus is a strong and ruthless defender of its own interests. Geneticists have graphically described how it is immovably plugged into its mother's metabolic resources and can manipulate them to its own advantage, just as some parasites can manipulate their victims. It can and does, for example, release some of its own hormones directly into her circulation, to ensure that the requisite amount of nutrition will be devoted to its needs, and not diverted to her own. After birth the relationship is transformed. The parasite-and-victim element has disappeared. There are now two separate sentient organisms, building up—gluing together—a social relationship: one which is, for all mammals, the forerunner and archetype of all ongoing social relationships.

At the outset, it is one of total dependency. It has been graphically expressed by saying "There is no such thing as 'a baby.' There is always 'a baby and someone.' " Usually it will be his mother who in most cases will have been successfully processed to love him and to be a soft touch. But even if he is orphaned, handed over to his grandmother or a nursery maid or an institution or adoptive parents, there will still be someone—or else very soon indeed there will be no baby.

When gluing together a relationship with a much stronger person, testosterone is not the hormone of choice to fuel your responses. Hawk-like strategies are inappropriate since you have no hope of influencing your opposite number by menaces. Babies, though they cannot hope to threaten their parents, are innately equipped with the power to evoke pleasure and empathy and pity. They are hard-wired to recognise human faces and smile at them, and we are hard-wired to be charmed by their well-known hallmarks—the big head, the big eyes, the chubby cheeks. Stephen Jay Gould once charted how cartoonmakers and advertisers cash in on these "Aw bless him!" triggers by endowing Donald Duck, for example, with chubby cheeks, even though real ducks have no cheeks at all.

Some people may impatiently protest that these comments are sentimental irrelevancies, that the behavioural hallmarks of infancy will all disappear without trace, like the milk teeth and the fontanel, and we will once more be brought up against the iron logic of the usual mantras: "Aggressive males father more babies" and "Nice guys finish last"—or at least it must have been so in the Pleistocene. So let us move a bit further along the male life cycle, and to remove any suspicion that the question is being contaminated by considerations of culture, or ethics, or political correctness, let us look at it in the context of our close cousins the chimpanzees.

Although the male infant chimpanzee may be able to use his charms on his mother, he will soon learn that he can't get around everybody like that: he must be prepared to stand up and fight for himself. And since that is inevitable, it is sometimes argued, the sooner the better. That is not quite true. In practise it should read: the later the better. He learns to socialise with other members of the band when he is still smaller than most of them. If his synapses are in good working order he will quickly find out—by trial and error, or by observing what happens to other uppity youngsters, or by looking to his mother for guidance—whose tail it is safe to pull, and who should be given a wide berth and treated with appeasement and subservience. His first really bitter conflict is liable to be with his mother when the time comes to wean him; since charm has failed him he falls back on rage and flies into a tantrum. Very exhausting, all that kicking and screaming—and it doesn't work. For the first 25% of his life span, though he may practise play-fighting with his peers, he will find it pays him best to keep his nose clean and his dove-like responses towards his elders and betters in good working order.

At adolescence, a change comes over him. It is not as comprehensive as the change that comes over a caterpillar or a tadpole, but it does mean that the final time switch in the y-chromosome sequence will be thrown, and his hormonal balance will go haywire. He may soon get involved in real fighting, the kind that may end in bloodshed, and he must get ready to engage in it. The sooner the better? No, the best plan is still: the later the better. If he picks a fight that he has a 50% chance of winning, he also has a 50% chance of losing, and the loss would cost him more than a few lacerations. The psychological effects can be crucial. Lose the first fight and you enter the next one with a doubt in your mind. Win it, your confidence surges and with it your testosterone. The trick is to play your cards right until you see the prospect of a contest that you have a 70% chance of winning. You may still lose. You may still be killed. 30% of male chimpanzee mortality is a result of these contests, and dead males who have over-rated their chances can do nothing to perpetuate their kind. The survivors who pass on their genes are likely to be those whose high testosterone and powerful muscles are strictly regulated by the synaptic talents that in humans we would call good judgement.

But the prize, we are told, is well worth laying your life on the line for. One day our hero may go for gold, challenge the alpha male, beat him in fair fight and become cock of the walk. Actually in real chimpanzee life, the palace coup is more often effected by a combination of bravery and diplomacy—two or three males forming a coalition. Even a temporary coalition means deploying some of the less hawk-like skills, a certain amount of restraint and mutual reassurance and solidarity. It also improves their chances if they can keep the females on their side as cheerleaders. The upshot will be that one of them succeeds to the top job. He may hold it for only a few years before becoming the ex-boss, and for the rest of his life span having to polish up the half-forgotten skills of how to get on with comrades that you cannot kick the tar out of. But while he is there, the story goes, he will have first choice of the best feeding places, and the most fertile and desirable females, and that will ensure that he begets most of the offspring. And from the Gene's point of view that is the only thing that counts.

For a long time that truth was held to be self-evident. Then it began to spring a leak. It started with birds. Studies making use of DNA analysis applied to white-crowned sparrows, and indigo buntings, and mallards, and acorn woodpeckers, all showed that between 25% and 40% of the chicks on the nests were the result of EPC (extra-pair copulation). Canadian scientist Lisle Gibbs checked this by monitoring in scrupulous detail the sex lives of the red-winged blackbirds nesting in a marsh near his university. DNA analysis showed that the number of offspring of a male blackbird bore *no relation* to the number of chicks hatched in the territory he so fiercely defended. Males with four chicks on their territories might have fathered five others elsewhere in the marsh, while males with ten chicks on their territories might actually have fathered only one. Marlene Zuk, describing these discoveries, commented: "It wasn't just a matter of a couple of chicks here and there. It was as if the entire method for calculating reproductive success, that cornerstone of evolution, was discovered to have a foundation of sand." The same thing is now known to occur in every avian family—"ducks, warblers, woodpeckers, wrens, orioles, the lot. This is that same group held up as a model of monogamy just a few short years ago. It was a real revolution and it took place within less than a decade."

But humans are primates and primates are a far cry from birds. In 1997 when it was decided to employ DNA analysis to establish paternity patterns within a troop of West African chimpanzees, nobody expected surprises on a similar scale. It was already well known that chimpanzees were promiscuous. The alpha male did not have exclusive access to females, only some degree of choice as to which female and when. The object was to try to quantify just *how much* the number of offspring produced by the alpha male was increased by his priority access to females at the peak of their fertility cycle.

The results were totally unexpected. The community being studied formed a discrete group with its territorial boundaries patrolled by a band of unrelated males. Sarah Blaffer Hrdy records that, "when Pascal Gagneux, David Woodruff, and Christophe Boesch analysed the genetic data, they found that just over half the infants born in this community (seven out of thirteen births) were sired by *outside* males. The fathers not only lived outside the study sample, but included males that the observers had never even seen the female travelling, much less mating with...Undetected by observers, female chimps were slipping away to solicit outsiders in spite of appalling risks."

Marlene Zuk sums up the conclusions to be drawn from the discoveries of the last decade. "Although in some species, the male best at brute-force combat can indeed control access to large groups of females, this is beginning to seem like an exception and not the rule. Males that appear to be dominant do not necessarily father more offspring, as DNA fingerprinting has revealed...The best we can do at this point is summarised in this statement from a book about animal conflict: 'In some primates, in some circumstances, dominant animals derive some benefit from their high status.'"

In short, the assumption that "more aggressive equals more offspring," which has long been treated as self-evident, has turned out to be untenable. There is no great mystery about why the genes of the nice guys were not wiped off the face of the earth. *Not* bidding for the top job, but just getting on with your life and not making waves, is a strategy that has much to commend it and does not necessarily involve dying without issue. It is true that men are up to a point genetically programmed to be more aggressive and more competitive than women. But the idea that for tens of millions of years natural selection has continued to keep up the pressure, in every generation inexorably selecting the most violent males to perpetuate their bloodline, has been discredited. The long-term trend is at least as likely to be in the opposite direction: selecting for increased ability to learn from experience, and increased cortical control over the promptings of our glandular secretions.

Is there any evidence that human males may have evolved away from aggression since they split from the apes? Two anatomical features seem to point in that direction: sexual dimorphism, and teeth. Men are on average bigger than women, but the difference is much less than in the other apes. Chimpanzee males weigh at least a third more than female ones, while a male gorilla is 50% heavier than his mate. Sexual dimorphism has unmistakably diminished in the human line over time. Also, male chimpanzees have long sabre-like canine teeth, which *Homo* has lost.

Richard Wrangham and Dale Peterson in their book *Demonic Males* suggested an explanation. They pointed out that apes and men can fight with their fists, and that the shrinking of the canines may well have coincided with the first appearance of tools and weapons. The weapons could have rendered the slashing teeth unnecessary, and may also have diminished the importance of bulk.

It is an ingenious argument, but not I think conclusive. There are other primates like muriqui monkeys which have also reduced their body size and their canine teeth to match those of the females, and they have neither learned to box nor invented the hand-axe. They are among the most peace-loving of the anthropoids, and it seems likely that the parallel changes in human males also represent a trend away from aggressive social interaction.

It is hard to assess the strength of the drive to physical violence in males if you are not one of their number. There seems to be a degree of peer pressure among them to lay claim to ungovernable surges of aggression, but you could have fooled me. I have moved around among the y-chromosomed creatures all my life and they seem well able as long as they're sober to get through the day (or the year, or the lifetime) without bashing anybody. I know there are parts of society where mayhem is the norm but off the screen I don't see it happening: how abnormal is this? I was reassured when the geneticist Steven Jones, who has written a whole book about the y chromosome, candidly confessed: "I have never punched anyone" (except once, apparently, in self-defence) and "I am as open to fantasies as the next man, but my dreams do not involve rape or torture." After reading William Hamilton's confessions about his murky subconscious, it was good to know that these visions are not obligatory.

Another cherished folk-scientific belief is that men are polygamous and women are monogamous. This is biologically nonsensical. Natural selection could never have moulded such an arrangement, destined to bring nothing but grief. Males are just as liable to fall in love as females are. All that love poetry was not just a literary convention. Apparently one of the effects of oxytocin is to induce females to become fixated on a particular male and feel that nobody else will do. In males, since they have a less plentiful supply of that hormone, nature has co-opted a different one called vasopressin—but it has precisely the same effect. The Don Juans of this world may be either deficient in vasopressin, or may be cases of arrested development, stuck in an adolescent phase of playing the field. The reasons why women were traditionally more desperate to marry and to cling on after the magic had faded were largely cultural and economic. Today they are often the first to want out.

We should not leave the subject of the *Homo sapiens* male without a final tribute. Down the centuries (as they are fond of reminding us) they have been responsible for most of the pinnacles of human cultural achievement in the creative arts and sciences and technology and exploration and learning in general.

Men of genius have changed our world and expanded our horizons.

We can point out with perfect truth that we were debarred from entering those fields; we were kept in bondage, and we carried the whole burden of bearing and raising the next generation, while they were free to indulge their curiosity and cultivate their minds. It's going to be different now. But it would be petty not to appreciate that some of them at least made excellent use of that freedom and produced some astonishing results. If they ever give up astonishing us, the world will be a poorer place.

> **Summary** *Human males are genetically predisposed to behave more aggressively than females. This difference has been culturally reinforced and exaggerated. It has tended to obscure the fact that they are also genetically predisposed to pursue their ends by means other than force whenever appropriate. For many of them "whenever appropriate" means "always."*

$$\overline{w}\Delta\overline{z} = \text{Cov}(w, z) = \beta_{wz} V_z$$

"What is love?"

Chapter 15
Right and Left

I think these arguments are pretty appalling. That they are so widely influential only shows how political questions about human evolution continue to be.

— Jerry Fodor

No one can make sense of the controversies surrounding mind brain genes and evolution without understanding their alignment with ancient political fault lines.

— Steven Pinker

These two quotations are from writers with very different views about life and the universe, but they agree on one thing: that the issues raised in the preceding chapters have become, once again, politically charged. If you know where an evolutionist stands on race, welfare, war and the environment, you can make a pretty good guess at where he stands on the New Synthesis.

The term "New Synthesis" (first used in 1942 about the fusing of the ideas of Charles Darwin and Gregor Mendel) has been revived to describe the attempt to merge the ideas of Charles Darwin and Alan Turing. It describes the perspective you arrive at if you follow all the signposts labelled Reciprocal Altruism, evolutionary psychology and reverse-engineered hereditary computational modules. One of its leading exponents has been Steven Pinker. Some people are disposed to believe him and others are disposed to doubt him, and as he suggests, the alignment tends to correspond to the viewpoints commonly described as Right and Left.

There is one school of thought which believes it is in your genes. This doctrine was immortalised in the last century by Gilbert and Sullivan:

> "Every boy and every gal
> That's born into this world alive
> Is either a little Liberal
> Or else a little Conservative."

It was recently restated in qualified form by Pinker: "Liberal and Conservative attitudes are largely, though far from completely, hereditable." In other words, you inherit a right-wing or left-wing personality profile which dictates your vote.

"Largely" hereditable is surely pushing it. At election times, political maps of Britain show bright blue Conservative areas in the prosperous south-east, while the rust belts left by decaying heavy industries to the north and west are bright red with Labour voters. That can hardly be due to random pockets of people with obstreperous genes. It suggests that your political leanings are largely, though far from completely, determined by the life you have led.

Pinker holds one other truth to be self-evident—that his own scientific views are not influenced either by his genes or his environment, but purely by his unrelenting search for the truth. This is, in a way, surprising because he once asserted that "With divisive moral issues, especially those on which Conservatives and Liberals disagree, all combatants are intuitively certain they are correct and that their opponents have ugly ulterior motives." No statement could be more lucid.

When did he decide that doesn't apply to himself? It happened when he made a delicious discovery. Not only is he intuitively certain he is correct, but in his own case, by pure coincidence, it just so happens that he *really is* correct. He keeps assuring us of this, as if he can hardly believe his luck. "As we shall see, the new sciences of human nature *really do* resonate with assumptions that historically were closer to the Right than the Left," and "My own view is that the new sciences of human nature *really do* vindicate some version of the Tragic Vision and undermine the Utopian outlook that, until recently, dominated large segments of intellectual life." Those are my italics I'm afraid, but you can hear them in your mind's ear as you read these passages. And after a heated BBC debate with psychologist Oliver James, he helpfully explained to a journalist where his opponent has erred. "James is a hard-line left-winger and he believes mine is a right-wing argument. I would say it is just sensible."

Another fault he finds with the Left is that they are "treacly" sentimentalists; intellectual wimps, not tough-minded enough to face the harsh truths about reality. They sometimes question whether all the truths about humanity have to be so harsh, so he undertook to present his own "distinctly unromantic theory of the emotions," revealing that they have a cold logic of their own.

His list of our coldly logical emotions is partly drawn from Trivers, and these are some of the definitions we are offered.

<u>Liking</u> is "roughly a willingness to offer someone a favor and is directed to those who appear willing to offer favors back."

<u>Gratitude</u> "calibrates the desire to reciprocate according to the costs and benefit of the original act."

<u>Sympathy</u> "may be an emotion for earning gratitude."

Happiness: Pinker defines this humorously with a quote from Ambrose Bierce, "an agreeable sensation arising from contemplating the misery of others."

Co-operation is defined as "an instinct to secure co-operation" as if co-operation was another word for obedience. That is nonsense; you cannot take the mutuality out of co-operation. It is a kind of grammatical error, like that of the woman who told her divorce lawyer, "*He's* the one that's incompatible. *I'm* not incompatible."

The need to mention parental love ties Pinker into knots. He grants it seven words. According to Hamiltonian kin selection it "ought to be vast and it is." At another point, he declares in the sardonic tones of one who has been goaded into repeating it, that parents are of course the most unselfish things *in the entire universe!* But he doesn't seem to have a clue as to what makes them tick. Twenty-odd pages on the parent-child relationship are full of parent-infant competition, infanticide, fratricide, cruel stepparents and cautions against believing in the more romantic accounts of bonding. Even the babies are deconstructed. A baby is apparently a master of hypocrisy; so ruthless about simulating misery to gain your attention that he will make himself really miserable just for spite ("self deception may begin early") thus explaining "the evolution of the brat." He even calls as witness a militant feminist, not I imagine one of his favourite sources of reference, to endorse that mothers often hate their children as well as loving them, because they are endlessly demanding and will suck you dry.

And what of compassionate married love and the love of friends—the stuff that suffereth long and is kind? For this we are trotted round the usual circuit of Reciprocal Altruism, complete with Prisoners' Dilemma and an extra twist supplied by Tooby and Cosmides, who call attention to an aspect of the logic of exchange they call the "Banker's Paradox."

So at long last we have an answer to the question posed long ago by William Shakespeare: "What is love? 'Tis not hereafter." "Nay, 'tis due to a computational module engineered to accommodate the Banker's Paradox aspect of the logic of exchange." As John Maynard Smith once mused about an EP exposition of mate choice in humans, "That is not how it feels."

Lefties, Pinker contends, are not merely sentimental, they are also terribly old-fashioned. He derides the way they all still believe in the blank slate theory. They all believe that there was a Golden Age when humanity was composed of noble savages. They believe that the mind is not a function of the brain, but is some mystic Cartesian entity called the "ghost in the machine." They have turned these out-dated principles into three sacred doctrines—empiricism, romanticism and dualism—which they are threatening to impose on everybody, and he fears they are winning. Presumably that is why he has entitled his chapter on the blank slate, "The Official Theory," without defining in what conceivable sense it is official. To him the threat is real and must be resisted at all costs.

Where does he find these people? I cannot think of a single living Darwinian scientist who subscribes to any of these propositions. Pinker offers us fourteen quotations in the course of four pages to prove that blank slate thinking is an ever-present menace. Eight of them date from the twenties and thirties of the last century. Only one is dated later than 1973. The most recent spokesman Pinker quotes as proclaiming his belief in the blank slate is Walt Disney. I must admit I had never thought of Walt as either a scientist or a radical.

One of his favourite examples of a dyed-in-the-wool radical is R. C. Lewontin, who not long ago restated his belief that "Variations among individuals within species are a unique consequence of both genes and the development environment in a constant interaction." That may sound as if he and Pinker are saying the same thing, but Pinker is not so easily fooled. Of course these people *deny* they believe in blank slates, but then they would, wouldn't they? Isn't there something deeply suspicious about the way they keep denying it? And yet, since they don't believe it, it is hard to see what else they could do.

There is something paranoid about this vision of a secret army of blank-slaters ready to wreak vengeance on anyone who casts doubt on the views of Rousseau or Descartes. "The radicals," he cries, "are now the establishment." He believes the world we now live in is as good as it gets and anyone who pretends it could be improved must be covertly out to destroy it. "Our first priority should be not to screw it up, because human nature always leaves us teetering on the brink of barbarism."

He has seen the barbarians in action. He writes of people being picketed, shouted down, subjected to searing invective in the press, even denounced in Congress, censored and assaulted and threatened with criminal prosecution.

What is being raised here is the spectre of Political Correctness, particularly in relation to academia. I somehow have the impression that he doesn't get around much outside that milieu. So when he says the radicals are now the establishment, he doesn't mean that George W. Bush and Donald Rumsfeld are barbarians in disguise, burrowing into the White House like termites to undermine the system. He means that the governing bodies of many American universities are wary of alienating potential sponsors and benefactors by appearing to condone the promotion of views by tenured professors which might be construed as racist.

In America political correctness is a serious problem and many people are alarmed by it. Philip Roth wrote a scary novel about it called *The Human Stain*, depicting it as a scourge of our time, a kind of mirror image of McCarthyism. His hero was an academic whose whole life was wrecked because, on a single occasion, he (improbably) used the word "spook" in its antiquated sense of ghost, and it was mistaken for a racist slur. The dice were further loaded by the fact that this ostensible Caucasian was in fact a black man in disguise, so no one could argue that he was anti-black, even on some deep subconscious level. Roth is a good enough writer to make me believe that there just might have been a guy like that, but he failed to make me believe in that whole campusful of hysterical youngsters, without a single second-year student who would have said, "Cool it guys, not that one. He's okay, he's been here for years; we know his track record." But perhaps I am being too sceptical. I couldn't have believed in McCarthy if he had been presented in the form of fiction.

Pinker is quite right to deplore the tactics of chanting and barracking and shouting down. They are mindless and self-defeating and designed to intimidate. They are of course by no means confined to the Left. Alfred Wegener got exactly the same treatment at Harvard when he stood up and tried to outline his theory of continental drift. The reference to assaults is even more serious, but he gives no details of this charge. Presumably he is not referring to the notorious glass of water. Yet that received such massive media coverage, I would have thought that if any professor had actually been punched that too would have gone down in history. He also implies that public expressions of political outrage are a growing problem. I am not so sure. The last instance of it he refers to was in response to the publication of *The Bell Curve* in 1994.

That episode is worth recalling because it was a textbook illustration of the mindset on both sides of the barricades. It was published, by R. Herrnstein and C. Murray, at a time when the political passions of the '70s appeared to have died down, and Murray explained, "If there was one objective that we shared from the beginning, it was to write a book that was relentlessly moderate in its tone, science and argumentation." It was accepted and published, and two weeks later it was praised in the *New York Times Book Review* as a serious work of scholarship. But the reactions that followed later were very angry. They were described as frenzied and hysterical in denial of the unwelcome facts revealed in the book—and above all as totally unforeseen.

The authors expressed their astonishment. Murray explains: "When we began work on the book, both of us assumed that it would provide evidence that would be more welcome to the Left than to the Right." Apparently they envisaged their left-wing colleagues coming up to them and thanking them for showing that the poor are congenitally dumber than the rich and blacks are congenitally dumber than whites, because that would strengthen the case for asking society to "compensate the less advantaged for the unfair allocation of intellectual gifts." Maybe calling for some kind of stupidity allowance?

What the Left reacted to was not the moderate tone of the volume, but the conclusion it reached. They perceived the moral to be: "There are too many blacks in this place, let's get them outta here." There is no nice way of saying that. "We feel that you might be happier somewhere else" has never taken the sting out of being fired. What the book did was amalgamate a scientific tone with a political objective. Segerstråle, who described the book as a "timely and well managed publication," also commented that unlike its predecessors, "this book did indeed vividly demonstrate the use of biological claims for legitimising a social state of affairs."

It is very easy to find evidence of a correlation between poverty and IQ. The only disagreement concerns whether poverty is the main cause of lower IQ or whether low IQ is the main cause of poverty. We know the arrow of cause and effect does not always flow from the genes to the eventual wage-bracket. We know that premature birth affects mental development and that higher birth weight is associated with higher IQ, even in identical twins. We know that prematurity and low birth weight correlate with maternal nutritional status, i.e. with poverty. After birth, we know that social background further influences development so strongly that adoption at birth of children from poor families raises their IQ by six points.

One significant aspect of *The Bell Curve*'s investigations is the number of aspects that fail to gain its attention. For example, for hundreds of years, the most glaring difference in congenital intelligence was held to be that between males and females. Eminent scientists assured us that, if women were rash enough to enter higher education, their little brains would become unhinged. Nurturists doubted that, and nurturists were right. Given equal opportunities, little girls grew up at least as good at passing exams as little boys. So in this book of over 800 pages, male-female comparisons get less than half a page. The comparisons are chiefly between different income levels and different races.

Within the racial context, some statistics are given far more attention than others. For example, according to the criteria used by *The Bell Curve*, Jewish IQ in America currently exceeds Christian IQ by almost as much as white IQ currently exceeds black IQ. But this analysis again gets only a fleeting mention. That is not what they want to talk about. They make no suggestion that this university could improve its intellectual standing by admitting that there are just too many gentiles around here.

Again, by the same method of calculation, Asians show up as consistently smarter than Caucasians. But here the authors are suddenly seized with doubts about the accuracy of their own criteria. They suspect the Asians aren't really smarter, they may be just showing off, "over-achievers," so no conclusions should be drawn. "On this issue," they report, "we will continue to hedge." Are you surprised? At another point, they find it sinister that black students sometimes get better examination results than white students *with the same IQ!* Could it be that, coming from a generally under-privileged minority, they too are overachieving? It has long been known in boxing circles that it pays to put your money on a hungry fighter. But in this connection, the possibility of over-achievement is not even glanced at. They conclude that somebody must be fiddling the results.

There is one other well established factor that affects IQ results, but is referred to only in a bracketed phrase within a subordinate clause of a single sentence, with no statistics attached to it. The sentence reads "The same factors that depress white scores (for example, coming from a rural area) will depress black scores." They don't want to touch that one with a barge pole. The urban-rural dimension is blatantly non-genetic. It might turn out to have a steeper curve than the black-white one and show urban blacks scoring higher than white hillbillies. And that, in turn, might raise queries about precisely what the IQ test is measuring. It may be culturally distorted to assess mental skills chiefly relevant to city dwellers.

All these and many other examples, without disproving any individual statement, make it clear that the book has an agenda. Consciously or unconsciously, it is cherry-picking its questions and tailoring its answers to serve that agenda. It is high-class propaganda, aimed at a highly educated audience, aware that such an audience may be disposed to overlook its tendentiousness if the aim is a noble one, namely to preserve high academic standards at all costs.

But is that really the aim? The book waxes indignant when some institutions of higher education offer financial incentives or ease conditions of entry in order to poach intelligent, black students from their competitors to make their ethnic profile look good. The only criterion, they seem to assert, should be intelligence. Yet they defend the easing of standards and the offer of generous scholarships to athletes who may be pretty dumb but will make their football teams look good. After all, that has always been done. It is a fine old tradition and there is nothing wrong with it. In short, the book is very strong on nostalgia. The authors would dearly like to get back to the way things used to be and are trying to find an oblique way of making that wish respectable. That is not a sin. In a free country, anyone is entitled to pursue a political objective and entitled to wrap it up in the vocabulary of scientific objectivity, just as anyone else is entitled to try to call their bluff.

But one side effect of the book was to debase the coinage of popular science writing. Up to that point books of this kind were mainly inspired by intellectual excitement over a scientific idea. That was 100% true, for example, of *The Selfish Gene*. Whatever was read into it, its purpose was never to tell people how they ought to behave. But since *The Bell Curve*, books about the biology of human nature have become increasingly prescriptive. It is now standard practise for New Synthesists to hang out a shingle, offering themselves as consultants for anyone concerned with administering social policies. The title given to the latest category of these consultants is "behavioural economists." Evolution and economics are getting increasingly intertwined. If it is true, as Pinker claims in a windy phrase, that "information is the lifeblood of the psyche," and if information comes in algorithmic lumps that can be fed into a computer, then who better than a Darwinist to hand out advice to economists?

The great thing about behavioural economics is that it is possible to write books imbued throughout with Pinker's Tragic Vision, and still come up with the happy ending that the unsophisticated reader craves. You can quote: "From the crooked timber of humanity no straight thing can be made," and assert that "our moral sentiments, no matter how beneficent, overlie a deeper bedrock of selfishness," and yet assure the reader that in the end all will be well.

The secret lies, as some of you may have guessed, in market forces. Matt Ridley, in *The Origins of Virtue*, waxes lyrical about them. "There is a bright side too," he reveals. "Its name is trade." He assures us that the magic of property turns sand into gold and that, as long as everyone selfishly pursues his own economic advantage, and as long as no stupid do-gooders try to interfere with them, order emerges perfectly out of chaos. He relates charming stories about trading practises in tribal societies: the Eskimos, fishermen on the Turkish coast, herdsmen in Turkana, villagers in northern India, the lobstermen of Maine, forest dwellers in New Guinea, and our own Pleistocene forebears in the Olduvai Gorge swapping reed baskets for bone hooks. As long as no do-gooders try to interfere with them, all works out for the best.

Private profit is the magic ingredient, and, if it goes for Turkish fishermen, Ridley assumes it also goes for Esso and ICI and Enron and Wall Street. They too represent private enterprise. Ridley warns that the actions of bureaucrats paid by the government may be distorted by self-interest and that will be very bad. But the wonderful news is that that doesn't apply to private enterprise. For those engaged in that, and the people paid by them, the more they are driven by their own self-interest, the better it works out for everybody.

In *The Origins of Virtue,* Ridley's own pet hates among naysayers were not the practitioners, nor the social scientists, but the environmentalists. They keep complaining about things like pollution and asking for it to be regulated by Governments. They are enraged, he says, when he assures them that "polluting companies adore regulation by Government," and that private enterprise is paradoxically the best friend that conservation ever had. Leave it to the multinationals. The world will be safe in their hands.

Steven Pinker puts the same faith in the magic wand. Since humans are wicked and stupid, they need "Systems that produce desirable outcomes, even when no member of the system is particularly wise or virtuous. Market economies, in this vision, accomplish that goal. . .it also follows that we should not aim to solve social problems, like crime or poverty, because in a world of competing individuals, one person's gain may be another person's loss."

These sentiments are in tune with the zeitgeist. The moral is that we should let well enough alone. Naturally the people who judge that things are well enough as they are, tend to be the people who thrive on the way things are. They feel comfortable with it. They can avert their eyes from the fact that the gulf between rich and poor continues to increase within most Western countries, and increases even faster between the richest and the poorest nations. They can always erect higher barriers between the socially excluded and the desirable residential areas. They can hire more guards, build more prisons, impose tighter curbs on immigration. They can demand an unimpeded passage of merchandise between states and a strictly impeded passage for personnel. They can buy up politicians, suborn officials, and make take-over bids for large chunks of the media. Meanwhile the Left is in disarray, disillusioned after the collapse of the Soviet experiment, and so far without an effective strategy for coping with the operations of a globalised economy.

Why then are people like Pinker raising alarm calls? Why is he summoning all hands on deck to repel the threat from the bunch of alleged blank-slaters that he claims are running the show? Perhaps the trust in the magic wand is not as absolute as it looks. The totalitarian Left once made the same mistake. They put their trust in Marx's magic wand, that which declared that there were iron laws of economics ensuring that history was on their side, regardless of the wickedness or stupidity of individuals. They too built more prisons and higher walls and censored inconvenient facts.

So the economists have come to feel a magnetic attraction to the biologists who tell them that *their* wand is far more magic and can never fail, because it is not only underwritten by Adam Smith and Malthus and Hayek, but also by Darwin, by the DNA in every cell of our bodies, and by the laws of human nature.

If they are reminded that absolute monarchy, slavery, war and patriarchy were also once said to be demonstrably rooted in our genes, they like to hear Pinker remind them why that kind of talk needn't worry them. "Those with the Tragic Vision are unmoved by ringing declarations attributed to the first person plural—we, our and us." In that vision, there is no such thing as "we": the genes speak the language of I, mine and me, and the scientificators speak as always with the passive voice.

The prophets of the right-wing renaissance give inspirational after-dinner pep talks to the weaker brethren, reminding them that greed is good. They should rejoice in it and boast about it. It is the lifeblood of the psyche that governs market forces. No more Mr Nice-Guy, their new guru, Roger Kimball tells them: "Benevolence is an instinct that should be subject to the greatest scrutiny. People have become virtue-intoxicated." This puerile do-gooding is spreading like bacteria, he warns, over the "rotting flesh of anxious bureaucracies, like the European Union, Oxbridge, the BBC and the United Nations," until the freedom of Western society and its very identity are under threat. It's designed to be scary stuff. It prompts the sentiment: "Joe McCarthy, thou shouldst be living at this hour." And bang on cue, a best-selling book has launched a campaign to dig up the late lamented senator and canonise him.

No sane person would suggest that the scientists of the New Synthesis are responsible for the politicians' drift to the Right, or vice-versa. It so happens that, at this time, they appear to give aid and comfort to one another. How much does it matter? Those people with a liberal personality profile, as we are asked to call it, won't be persuaded to seek a cure for their virtue-intoxication. They won't follow Kimball's lead by repeating to themselves: "Every day and in every way we are getting nastier and nastier." They will continue to think that there are imperfections in the way the world is managed, and that with good will and co-operation, improvements can be made.

Nevertheless, the tide at present seems to be flowing against them. An indication of the state of play can be gleaned from lists of publications over the last few years, and sometimes even from their titles. One recap of New Synthesis ideas was entitled *The Triumph of Sociobiology*, while the most comprehensive collection of arguments questioning behavioural psychology, edited by Hilary and Steven Rose, was called *Alas, Poor Darwin*. The danger is that the Left's morale could be undermined by the constant drip-feed of proclamations that science itself is against them, because it proves that what they hope to do is demonstrably impossible.

Fortunately, that is not the case.

> **Summary** *Science aspires to be uncontaminated by value judgements and for the most part, it succeeds. But that is not true of the channels by which its discoveries are filtered through to the outside world. That particularly applies to the sciences that investigate the nature of human beings.*

"..heaven's cherubim horsed Upon the sightless couriers of the air."

Chapter 16

Striding the Blast

To appreciate what has happened, you will have to abandon cherished notions and open your mind. You will have to enter a world where genes are not puppet masters pulling the strings of your behaviour, but are puppets at the mercy of behaviour.

— Matt Ridley

The process of lifting the lid on the genes continues to gather pace, but the results have not been quite as anticipated. A single gene was found to be involved in the inheritance of cystic fibrosis, and another one for Huntington's chorea. These were seen as the forerunners of a host of others, and we seemed to be on the brink of a bright new world. Just as scientists had found "the germ for" an infectious disease and then ways of curing or preventing it, they would now find "the gene for" a non-infectious disease and shortly afterwards a cure for it. Perhaps they would also find the genes for crime and sexual deviance and drug addiction and cure those too.

Those hopes were short-lived. Nowadays many scientists wince on hearing the phrase "a gene for. . ." It is not that simple. They have learned that different conditions in different people may be caused by the same gene, and the same condition in different people may be caused by different genes. They know it takes a combination of 39 genes to determine the colour of a fruit-fly's eye. A great deal of effort went into the search for the genetic cause of schizophrenia, but it petered out after the condition had been linked to markers on nearly all of the human chromosomes. "Only six chromosomes (3, 7, 12, 17 and 21) do not have putative links to schizophrenia, but few of the links prove durable." They now think in terms of sets of genes, combinations of genes, or of "a myriad of interacting genes." Genes interact not only with one another but with the organism and the world outside with which the organism has to cope.

This aspect of how the gene works is not a new discovery. Scientists never did talk of a gene causing a particular behaviour and if they did, as E. O. Wilson assures us, they never meant it literally. It has long been known that the effect of a gene depends on the organism's environment, even among plants. Genetically identical seeds of the arrowleaf plant may develop into one of three strikingly different shapes and sizes, according to what kind of habitat the seed falls into. But the extent to which the environment modifies the action of the genes is much greater than had been realised. It has called for what in political terms might be called an agonising reappraisal, as indicated by the quote from Matt Ridley at the head of this chapter.

His latest book, *Nature via Nurture*, gives a vivid account of the change, updating and in some respects, subverting the views he propounded in the *The Origins of Virtue*. He has seen a new light and this time he needs no excursion into economics to provide his happy ending; for he is reporting a new consensus, a new consilience. "Nature versus nurture is dead. Long live nature via nurture." Both sides in the debate have been vindicated; they were all really saying (or at least meaning) the same thing. He assures the nurturists that the genes were really on their side all along: "Genes are the very servants of experience."

He would not be human if there weren't one or two cherished notions that he cannot quite bring himself to abandon. He still, for instance, invites us back to the Pleistocene to meet some hunter-gatherers called Og, Iz, and Ob, and he still thinks that the magic hand of the market is the only thing that can lead us into a better world. But he gives an excellent and readable account of the new world he invites us to enter. We hear less about the computational elements in our brains and more about hormones. There is less of the crystalline clarity of the algorithm and a greater readiness to start again from scratch. It is reminiscent of T. H. Huxley's exhortation to sit down before truth as a little child, even those aspects of the truth that cannot be mathematically quantified. Perhaps for the time being we have exhausted the novelty and the usefulness of peering at life exclusively through the gene's eye lens. It might be worth while to open the other eye again and remind ourselves of how the landscape used to look before we wandered down that particular path.

One useful reminder lies in the term instinct. It used to be a standard term, even an indispensable one, in discussions of animal behaviour, but in recent years it has tended to be avoided. For some scientists, it joined phlogiston and protoplasm as a shibboleth for identifying people who have not kept up with the times and are not worth arguing with. The only reason given for avoiding it was that it is hard to define it with precision, but that is also true of some of the concepts of the New Synthesis writers. The taboo against it has never been watertight; it slips in and out of the vocabulary of many of the New Synthesis writers in a casual manner. Pinker casts a glance at the traditional four F's and he suspects that we may have more than four instincts. Ridley believes we have social instincts. I agree with him.

But when they do venture to utter the phrase "social instincts" they use a quite different vocabulary to deal with it. These instincts it seems are not on the same level as the great dark world-shaking passions like fear and hate and wrath and jealousy and sadism. They consist of a clutch of cool calculated behaviour patterns. They sound more like acquired or imposed behaviour than anything more deep-rooted. They include the ability to "learn how to co-operate; to discriminate the trustworthy from the treacherous ; to commit themselves to be trustworthy; to earn reputations; to exchange goods and information and to divide labour." You never hear of people learning how to betray, or to exchange blows, or committing themselves to be cruel.

I do not recognise this portrait of my species. I hope I may be forgiven, in the name of consilience, if I call on one of the big guns from the other side of No Man's Land. Nobody I think would call Shakespeare a sissy. He has gone eyeball to eyeball with cruelties and treacheries and horrors. He has portrayed murders galore, suicides and tortures and mutilations; hypocrisy and greed; motiveless malignancy; the slaughter of innocents and the descent into madness. But he knows that is only half of the picture. Take just one of the four-letter words that fail to make it to Pinker's list of emotions—pity. It is not included there because no amount of ingenuity can reverse-engineer it into a disguised form of reciprocity, since it is selectively bestowed on those least likely to be able to reciprocate.

In Shakespeare's plays it can move mountains. Macbeth knows in advance that if he kills Duncan, it is a force that can bring him down.

> "And pity, like a naked new-born babe,
> Striding the blast, or heaven's cherubim, hors'd
> Upon the sightless couriers of the air,
> Shall blow the horrid deed in every eye,
> That tears shall drown the wind."

Mark Anthony too knows better than to try to rouse the populace by appealing to fear or trying to smear Brutus. He reaches for a stronger trump card. "If you have tears, prepare to shed them now."

Conjuring up such tidal emotions is called demagogy if you don't share its aims. If you do, it is called heroic oratory, like Henry V's exhortation to "close the wall up with our English dead." It is dangerous stuff. The point I am making is that it only works, for good or ill, if it makes contact with something that is primal and basic in the human heart and not just a mental slide rule, prudently calculating the odds on getting a pay-off. It may be objected that Shakespeare's Antony and his Roman rabble were figments of a playwright's romantic imagination and that in real life, pity never actually strides the blast. Except that sometimes it does.

Consider Britain's abolition of the slave trade at the beginning of the 19th century. What in the name of evo/psych was going on there? It had been in every way in the country's interests for that respectable mercantile operation to be sustained. A class of entrepreneurs had perceived a gap in the market and was catering for it; it was earning them vast fortunes. In ports like Bristol, the wealth was trickling down. The merchants built fine mansions, raised cultured families, paid their taxes, went to church on Sundays and patronised the arts. They incidentally helped to ensure the supply of cheap sugar from Jamaica to the great benefit of the balance of trade. The magic hand of market forces ensured that everybody benefited, except of course the raw material, the living freight their vessels carried. That cargo was liable to a regrettable percentage of wastage en route, but not more than the market could stand.

In 1787, a handful of naysayers like William Wilberforce expressed outrage and undertook to put an end to it. There was only one emotion they could appeal to: "If you have tears, prepare to shed them now." At first, they were laughed off. No one in Britain ever even set eyes on the merchandise; all that was happening in distant continents, in Africa and across the Atlantic, and not to people like us. But here and there in the fine mansions, the darling daughters at the dinner table were beginning to look troubled. "Papa, can it really be true that in those ships. . . ?" And within twenty years, the hard-nosed politicians calculated that it would pay them best to ride the wave of compassion, rather than trying to resist it.

Recalling that episode is not an attempt to get treacly about our species or uncover a noble savage at the heart of it. Ending the slave trade was one illustration of human nature in action; the holocaust was another. It simply reinforces a running theme. Human nature does not consist of powerful, basic, biological drives to be selfish and cruel, tempered by well meant, cultural attempts to be prudent and restrain them. The instinct to be humane is as deep and Darwinian and irrational as any of the others. There are conditions under which it is not activated, but that is equally true of the more destructive instincts like aggression.

There is unanimity on one point. We are a social species. There must have been an adaptive advantage in living together in groups. If it was adaptive, we would have acquired some instincts inducing us to live together. A sixth F, perhaps? F for fellowship? One approach was to study other social species and look for analogies with the evolved structures of human social life and the roles played by individual members of it.

It did not prove a very fruitful method. In many assemblies of animals there are no roles. They mass together in vast numbers but their behaviour is unstructured. This tends to happen where food is virtually unlimited, as with vast shoals of fish feeding on plankton and vast herds of wildebeest feeding on grass. It may look as if the shoal of fish is faithfully following the one in front which presumably knows how to find the best food sources or how to escape predators. But in the sea the plankton is everywhere, and if the shoal is being chased, the one in front would be the last to become aware of it. The gene for shoaling issues only one instruction: "Try to get behind somebody." The fish in front, it has been discovered, is the one with a slight defect, rendering it the least successful in obeying that order. We can't deduce from this that our own societies are led by individuals who are not quite all there, though on occasion it may seem like that. But it illustrates how careful you have to be with these comparisons.

One comparison that misled the Victorians was that of the beehive. Some of them regarded it as the model for a successful society, with one individual at the top like their own dear Queen and all the others contentedly knowing their place—long live hereditary monarchy. They were missing the point that the egg from which the queen bee emerged was genetically identical with the eggs from which the worker bees emerged. The difference lay in the upbringing of the larva which the workers selected for special treatment. That is a far cry from democracy but as a way of running a society, it is more like the Dalai Lama than the Prince of Wales.

Researchers might once have hoped that by narrowing down the search to our nearest biological relations, a clearer picture would emerge, but they would have been disappointed. Societies of monkeys and apes come in all sizes and varieties. Some live in very large bands like the grass-eating gelada. At the opposite end of the spectrum, the gibbon picks a mate for life and they then live together at the greatest possible remove from all others of their kind, even from their own offspring as soon as they are old enough to be chased away. Some baboons live in harems with a fascistic macho male terrorising his subordinates into obedience, while the bonobos have no male supremacy and no hierarchy; they quickly apologise if they have upset anybody and believe in making love not war. Their relationship to us is exactly as close as that of the chimpanzees who are always squabbling and forming short-lived coalitions and hoping to pull off a coup.

One factor that humans share with all apes and monkeys is that their social behaviour is at least partly cultural, determined by experience and reliant on imitation. There was evidence for this in Harlow's discovery that in monkeys, even the ability to copulate depends on having seen it done. There was further evidence of it in Frans de Waal's experiment involving two species of monkeys—rhesus monkeys which have a strictly enforced hierarchy, and stump-tail macaques, which have a much richer repertoire of gestures of reassurance and reconciliation and are three times more likely to use them to establish good relations after every dispute.

The method used was to establish mixed species of groups of two-year-old rhesus monkeys and slightly older stump-tails and leave them together for five consecutive months. Whenever a rhesus monkey approached a stump-tail with a grunt of challenge, instead of the challenge leading to a counter threat, a chase or a fight, the stump-tails did not even look up. In the course of the experiment, de Waal notes, the rhesus monkeys learned this lesson a thousand times over. Physical violence and injuries became virtually absent, friendly contact and play became the dominant activity. The rhesus's rate of peace-making after fights grew until they reconciled exactly as often as the stump-tails. Even after the experiment, when the rhesus monkeys were left to interact among themselves, they "maintained this newly acquired pacifism."

Here too, the moral is not necessarily a sentimental one. If the experiment were reversed, the younger stump-tails being apprenticed to older rhesus ones, they might have closed the cultural gap by becoming more aggressive. But it is further striking proof of the importance of the learned element in apparently instinctive behaviour. To those who says you can't change human nature, one answer would be that you can change even monkey nature, as de Waal did, without too much difficulty and without any Pavlov-style conditioning by human beings.

John Maynard Smith clarified the issue. "These two kinds of behaviour, instinctive and learned, are not sharply distinct but occur side by side . . .The learned component is increased in the social primates and vastly increased in *Homo sapiens*." He concluded that "in man, the instinctive component is difficult to recognise. There is no pattern of behaviour more complicated than the sucking of a baby which does not require to be learnt."

Any political deductions from the biological data (and this is the dimension we are trying to investigate) have to be based on this agreed fact. The two components exist side by side. On a philosophical level, they are equally valid and important but if you want to change society for the better, you need to decide which component you can most easily influence: the inheritance or the environment.

There is no contest. If it were to be proved that some element of criminal behaviour is rooted in the genes, the only way to influence that element, short of sterilisation, would be by gene therapy for behavioural malfunction becoming a practical option. It is very doubtful that any society could or would afford to carry out that operation on all the inmates of its prisons, and then let them go free. The other way science could help is to find out in what conditions the potentially criminal genes are *likely to be expressed,* so that we could try to change those conditions. But that involves precisely the kind of statistical information which evo/psych researchers routinely screen out. Their work is philosophically valid but they should not be surprised when it is treated as politically unhelpful.

In Ridley's new interactive world of nature via nurture, how does he account for our acquired ability to co-operate with one another? It must have come from somewhere. Evolution never starts from scratch; it works by gradual modifications of something that had already existed. Ridley offers us the old answer, a barely modified version of Lorenz's statement that love is a side-effect of aggression. "The co-operation," writes Ridley, "is in the context of competition and aggression."

That is part of the picture, the androcentric part, and it refers to male bonding. The greatest heroism, the fiercest loyalties, the most valorous deeds of self-sacrifice are displayed when males combine against a dangerous common enemy. The horrors of war cannot detract from the courage it evokes or our admiration of that courage. This kind of bonding was almost certainly selected for primarily in males but is now part of the human heritage. While war is on, it engulfs most people. Women tend to identify with the solidarity and the willingness to endure hardships. They may volunteer to join the armed forces. In occupied countries they may go underground with the freedom fighters. And they endorse the conviction that the enemy is evil. In wartime that goes without saying.

But most people do not live all their lives in this context and societies do not fall apart when the war is over. Something still makes it possible for people to live together and co-operate in the reconstruction, rebuilding the homes, growing the food, raising the families, nursing the wounded back to health. Some part of that caring was almost certainly selected for primarily in females but it too is now part of the human heritage. In the field hospitals of the two World Wars, as in the fictional MASH compounds in Korea, there were male doctors and orderlies as well as nurses who cared for any casualty that was brought in, without asking whether or not he was one of the evil enemy. In 1914, the poets of England were saying that the poetry of war was in the glorious bravery of those who poured out the red sweet wine of youth. By 1918, they were saying with Wilfred Owen: "The poetry is in the pity."

Any "genes for" social cohesion are likely to be derived from more ancient instincts, originating perhaps partly in male and partly in female behaviour patterns. The way in which they are expressed may differ according to whether they find themselves in the male or female organism. It's now increasingly recognised that the way they are expressed may also differ according to the habits, predicament, and life experiences of that organism.

That plasticity is basic. The ambivalence applies even to hormones. The hormone oxytocin, which in almost all situations serves to damp down aggression, has the opposite effect in female animals when there is a threat to their young. It prompts the reaction which gave rise to the dictum about the female of the species being more deadly than the male. We might conclude then that in conditions of intense competition, social cohesion derives in a greater degree from tribal solidarity. In the absence of such conditions, it derives in a greater degree from empathy. If that is true, it is worth bearing in mind.

Some people are at a loss to understand why left-wing thinkers make a big issue out of such questions. Being in thrall to your environment is no more liberating than being in thrall to your genes. If you were born in the wrong place at the wrong time, or suffered traumas in your early years, that can mess up your life just as irrevocably as being born with a hereditary handicap. It is no comfort to be able to blame it on circumstances rather than heredity. That only introduces a different kind of determinism. So what's the difference?

The difference is this. There is one human characteristic that we share with no other animal. We are able to envisage events that will occur when we are no longer around to see them. We can not only envisage them but care about them. It makes a powerful difference to the quality of life here and now if we have reason to believe that things are getting better. Our children (or if we have none, other peoples' children) will have a better life than we have had.

A great deal of human effort is invested in projects that will not come to fruition in the lifetime of those expending the effort. It is not a reciprocal process. They don't stop to figure out: "Wait a minute, what has posterity ever done for me?" They plant orchards when they won't be around to eat the fruit. They devote their lives to conservation schemes so that tomorrow's children will still be able to see tomorrow's pandas and rhinos. They have their names carved on monuments and park benches in the hope that people yet unborn will read them. They worry about the results of global warming, even though they may be safely out of this world before the consequences of it become seriously uncomfortable.

Everyone displays this tendency in some degree. Up to now, it seems to have loomed larger in the consciousness of those who have had a raw deal in life, and of those who empathise with them. People belonging to classes which were formerly in the ascendant have been more likely to look backwards and dwell on the past. It was the proletariat (that's derived from a word meaning "offspring") which was energised by dreaming dreams of a better future. "Come the revolution . . ." They believed in something called progress.

Darwinists stress that in biology there is no such thing as progress, only change. But for most people that idea is hard to assimilate. We are clearly different from animals. We feel as if the difference between us and the other apes deserves to be called progress. However passionately we disclaim any idea of being made of different stuff, or being in any way morally worthier, we are certainly much more competent. We are capable of cherishing an inner conviction that problems are there to be solved.

When we read of *Homo sapiens* getting stuck for thousands of years on an unchanged design for the Acheulean Axe, we are tempted to feel a touch of puzzled impatience. What kept him?

But recently the feeling that changes were not coming fast enough is being overtaken by the fear that they are coming too fast, and that human nature, being still stuck in the Pleistocene, may be unable to cope with them. It is the mood that Rabbie Burns was in when he wrote:

> "But forward, though I cannae see,
> I guess, and fear."

Summary *The explosive expansion in the genetic sciences encouraged the belief that the genes were the prime determinants of human behaviour. That was never quite what the scientists said. As their understanding continues to deepen, they are ever less likely to say it. Some of the thinking that preceded the explosion deserves to be re-evaluated.*

This instinct...described as male bonding.

Chapter 17
Progress

All corporeal and mental endowments will tend to progress towards perfection.

— Charles Darwin

That statement appeared on the last page of *The Origin of Species* and it justified the upbeat vision of evolution which T. H. Huxley retailed to his eager audiences. It may be that reaching the end of the immense labour entailed in writing his book had raised Darwin's spirits to an uncharacteristic level. In an earlier chapter, he had offered one reason for believing in progress but the tone had been far more cautious: "The inhabitants of each successive period in the world's history have beaten their predecessors in the race for life. . .and this may account for that vague and ill-defined sentiment that organisation, on the whole, has progressed."

Words like vague and ill-defined suggest that his faith in progress was never very robust and by 1872, it had vanished altogether. In a letter to an American paleontologist he wrote, "After long reflection I cannot avoid the conviction that no innate tendency to progressive development exists."

Yet most philosophers in Darwin's lifetime strongly believed in progress. They saw people all around them busily engaged in increasing man's control over nature, building railways and schools and hospitals, improving sanitation, adding to scientific knowledge, extending the franchise, introducing more humane conditions into prisons and asylums, and they couldn't help feeling that things were getting better all the time. They didn't deny the existence of injustice, greed and corruption but could regard them as regrettable defections from a decent human norm.

They were over-sanguine. Today, after the bloodiest century in the history of humankind, there has been a massive mood swing. Anything evil in human nature is widely regarded as endemic and destined to prevail. Some even declare it would be better if the rapacious biped, *Homo sapiens*, had never come into existence—and this is treated as a profound truth, spoken with becoming humility. We set progressively higher standards for what we regard as civilised behaviour, and that's fine. We then descend into bitterness and self-flagellation whenever we fall short of attaining those standards. And that's merely debilitating.

This change in the zeitgeist is more unwelcome to the Left than to the Right. People who aspired to reform society were strengthened by the feeling that history was on their side. They could only be daunted to hear that this is not the case and things can go disastrously wrong. Those who are doing well out of the status quo have more incentive to boost the dread of change with Belloc's warning, "Always keep a hold of nurse for fear of finding something worse."

So it is ironic that Stephen Jay Gould, in his youth a Marxist activist and the white hope of the Left, was in later life an insistent campaigner against "the errors of progressivist bias." His influence was immense, but every guru is fated to find that each year brings new ranks of eager acolytes asking questions he finds unutterably boring, because he had answered them so long ago and so many times. Gould acquired the habit of issuing ex-cathedra statements, known in scientific circles as the JPW response: That idea is Just Plain Wrong. Many students are happy to be told what they ought to think, but his peers were less impressed and apt to JPW right back at him.

It is puzzling that he expended so much energy on the progress issue. As with many arguments about evolution, the basic facts are not in dispute: it is how they are interpreted. In his early days, he had been happy to explain why it was advantageous for the panda to acquire a fake thumb and for the flamingo's smile to be upside down. But later, he came to feel that this kind of explanation, implying that every change in any species must be adaptive, was being carried to absurd lengths. In a few instances, it was. He rightly stressed that evolutionary options are limited by numerous restraints; natural selection is a process of make-do and mend, not a way of designing perfection. He avoided the word progress in favour of the less emotive "directional variability."

So far so reasonable. But he went further. He implied (rather like Pinker with his Blank-Slaters), that there was an Adaptationist Programme, characterised by a deluded and clap-happy ("Panglossian") faith in perfectibility, tainted with "Californian touchy-feelydom." He attributed it to a wish to keep the human race on a pedestal of arrogance. He countered it by stressing the importance of contingency, especially in the form of cosmic disasters and mass extinctions. By the time he had finished, progress was not merely an illusion, it came close to being a dirty word.

Resistance to this view came from, among others, Richard Dawkins. That was not quite as surprising as it sounds, for his books had always contained some upbeat passages. *The Selfish Gene* described the evolutionary trend to increased intelligence in humans and envisaged it as an ongoing process: "But as brains became more highly developed, they took over more and more of the actual policy decisions. . .The logical conclusion to this trend, not yet reached in any species, would be for the genes to give the survival machine a single, overall, policy instruction: do whatever you think best to keep us alive."

If you believe that the gene is selfish, you should rejoice that its power to give policy decisions is being progressively curtailed. If you believe that stupidity can mess up human lives even more effectively than malice, it is encouraging to think that the development of intelligence may be proceeding towards a logical conclusion. The passage would also explain why, in Steve Jones's phrase, we can already tell our genes to go jump in the lake if we don't like the advice they are offering us.

In *A Devil's Chaplain*, Dawkins pointed out that the argument about progress was a semantic one: it all depends on what you mean by progress. Gould defined it as a movement towards an ultimate pre-existing ideal and found that version easy to knock down. Dawkins opted for a more realistic definition: a tendency for lineages to keep on steadily improving their adaptive fitness to the circumstances in which they find themselves and "By this definition, adaptive evolution is not just incidentally progressive, it is deeply, dyed-in-the-wool, indispensably progressive."

Meanwhile, new voices were being heard. Advances in technology, which had already transformed the work of geneticists, were beginning to transform the work of anatomists and they were using these new techniques to enable them to talk in new ways about the emotions. At one time, the chance of reaching any scientific conclusion about the way people and animals feel was close to zero. Animals can't tell you how they feel; people will talk about it but are liable to exaggerate or lie or deceive themselves. So psychological theories about the emotions had been heavily dependent on the imagination.

There was, for example, a Hydraulic Model which owed much to the age of steam, envisaging our psyches as cauldrons of pressurised emotions needing to be channelled, sublimated, vented, discharged and provided with outlets. This metaphor was particularly prevalent in the discussion of sexual emotions and had clearly originated in the brain of a human with an XY chromosome, and hence a penis. Other models envisaged the psyche in layers, upper and lower, conscious and subconscious, or divided into three—id, ego, super-ego. If this was science, it was very soft science; soft to the point of deliquescence.

Today, discussions of the emotions are no longer dependent on metaphorical constructs. They are based on detailed examination of glands and their secretions, on chemical analyses of hormones, on experiments to show how their presence or absence affects animal behaviour, on identifying and counting the number of receptors in the brain designed to receive messages from the various neuro-transmitters, and on brain scans involving magnetic resonance imaging.

These scans can identify pleasure areas and pain areas in the brain and render them visible on a screen. Contemplation of primal urges and the Oedipus complex is being replaced by research into seratonin and dopamine, the anterior cingulate cortex and the right ventral pre-frontal cortex. All this is still in the early stages, but it is growing as fast as genetics and is transforming approaches to the question of why we behave as we do.

We may as well stick with the useful image of a purposeful gene which has by now firmly taken root in the public mind.

How does the gene get its desires translated from DNA into social behaviour? In the prevalent model, it sends a message direct to a computational brain module which instantly flicks through its files, confirms that person B has not defaulted on the repayment of favours in any previous exchange, and then instructs person A that it's okay to be nice to B. The interpersonal emotions are, as it were, filtered through the cognitive faculty and licensed to exist.

The emotional model is different. It suggests that the gene uses a carrot and stick method. It installs pleasure areas and pain areas in the brain, designed to serve its overall aim of "keeping us alive." It ensures that an organism will eat, by causing hunger to hurt and eating to be enjoyable. In many species, this works so well that cognitive areas are not needed and are barely discernible.

According to this model, in animals with more complicated problems to solve, reason evolved to subserve the emotions. Reason cannot initiate action because it cannot want anything. It can only advise us how best to attain our desires. When we have conflicting desires—i.e. most of the time—it can arrange them in order of priority. It can weigh long-term aims against more immediate ones. (Being fat later will hurt more than being hungry now.) "I feel. Therefore I think."

Can we assume that emotions are moulded by natural selection to fit the needs of the organism? Yes, certainly. The fight/flight reaction is adaptive in most animals, triggered by loud noises or large objects approaching at speed. It would be surprising if it played much part in the emotional repertoire of Galapagos tortoises since there is nothing that threatens them which they can outrun. Humans still experience vertigo, a reaction to heights adaptive in our arboreal ancestors; it would be pointless for dolphins to experience vertigo because dolphins don't fall down.

Can we further assume that the five million year period, since we shared a common ancestor with the apes, is long enough for us to have evolved a different set of emotions? Yes, certainly. The monogamous and promiscuous strains of the prairie vole are as closely related as men and apes, but have evolved different emotions. That was evident from their behaviour and is now detectable in their brain centres.

Does this mean that we have an emotional repertoire different from that of the apes? The answer is yes. Ridley, in his last book, devoted a few pages to "the oxytocin story" and made what he seemed to think was a daring prediction. He thought it might one day be discovered that apes "have fewer oxytocin receptors in their brains than human beings." I am convinced he is right.

There used to be an unspoken assumption that, while cognition evolves, emotions remain static. It was implied that we needed stronger reasoning powers, at least in part, just because our emotions had remained primitive and unreconstructed and so badly in need of curbing. Now there is a growing recognition that emotions too are selected for. But there is a curious reluctance to take the obvious Darwinian next step and pose the question that is crying out for an answer, the kind of question that is routinely asked about our physical attributes. Have we evolved to become nicer than the apes and if so, how much nicer and why?

It's a valid question but too provocatively phrased. The problem is not merely that words like nice and loving are imprecise (they are not more imprecise than other layman's words like hate and greed) but they do uniquely provoke in many scientists a squeamishness that positively makes their toes curl. And we don't want that. "Eusocial" is a good word. It refers to anything that fosters co-operation between individual organisms rather than conflict. The proposition then is that *Homo sapiens* has emotionally evolved to become progressively more eusocial than the apes. For all I know that may be a truism which everyone tacitly accepts, but I do know that it is rarely, if ever, proclaimed. I would like to proclaim it, if only in order that it may be rationally contested and shot down if it is wrong, rather than treated as touchy-feely and speciesist and taboo.

A change in our emotions has not consisted of anger, lust and aggression being selected out or weakened: they are still there in force. Compared to animals, humans are better able to control them because they are often counterproductive in the long run and humans are usually, at least when sober, reasonable enough to figure that out. What is new is that in humans the gamut of eusocial emotions such as pity (discussed in the last chapter) has become more varied and stronger. Like other evolutionary changes, the innovations have not appeared overnight but have arisen out of continual small modifications of features that already existed and are shared with other animals.

Two of these are easy to identify. One is kin selection. Its primary manifestation is parental care. It is obvious that this would need to be augmented as human infants were delivered at a progressively earlier stage of development, and remained helpless for an increasingly long period after birth. The gene urgently required parents, particularly women, to respond more unselfishly to the spectacle of infantile helplessness and neediness. It was crucial to species survival. If it sometimes happened that helplessness and neediness in others pressed some of the same buttons and resulted in "non-adaptive" forms of kindness, then from the gene's point of view, that was a price well worth paying.

Non-parental aspects of kin selection may also have become blurred at the edges as our ancestors became unable to identify their blood relations with any certainty. Most other species can perceive olfactory resemblances between individuals at least as readily as we use visual resemblances, and use them to recognise their own kin. It has recently been confirmed that a male baboon can distinguish his own offspring when he encounters them simply by the smell of them. This discredits all those fantasies about females being promiscuous so that numerous males will remember their brief sexual encounters months later and each will assume that her baby may be his child.

But human powers of olfaction have diminished so far that we cannot know at first hand who is a sibling and who is not. We only know which children we were brought up with and what we are told about their parentage. In Israel, when children were brought up on a kibbutz, the instinct not to mate with siblings made them averse to sexual relations with any member of the opposite sex reared in the same kibbutz. Such uncertainty could mean that emotions which originally evolved to implement pure kin selection could spill over, more readily than in any other species, into relationships with non-kin. A "fraternal" degree of goodwill can thus grow up between buddies who live or work in proximity, even if they are genetically not brothers or even cousins.

In many social species, there is also an ad hoc kind of brotherhood which operates when a group is confronted with a challenge. Adults, males in particular, will put any existing rivalries and frictions on hold and co-operate for the duration of the emergency. This instinct, described as male bonding, enables them to work together in hunting prey, as wolves and dolphins do, or in defending their territory against other bands of the same species. Within the band, it fosters the capacity to display loyalty—and that too is a eusocial instinct capable of evolving into something stronger and more permanent. Unfortunately, as compared to the goodwill derived from other biological roots, this one, being testosterone-based and adrenaline-fuelled, has a dangerous aspect. It needs an enemy, real or imagined, to activate it and sustain it.

An indication of progressive adaptation to social life in humans is our increased sensitivity to the emotional state of other people. Eye contact between animals in the wild is rare and fleeting, except in highly charged, eyeball to eyeball confrontations signalling imminent aggressive action. Meerkats, for example, are highly social animals but you never see them gazing into one another's eyes. Humans, when in company, are constantly monitoring the emotional states of their companions, alert to mood changes signalled by tone of voice or facial expression. Between people who know each other well such changes, even subtle ones, are virtually impossible to hide. They prompt the instant question, "What's wrong?"—and the answer "Nothing" never convinces.

Again, the evidence for this is more than just a gut feeling. Darwin wrote an entire book about the expression of the emotions in man and animals. Some human expressions (raising the eyebrows or a curl of the lip) have parallels in animals, but our repertoire of such signals has dramatically expanded. They are instantly recognised in human communities all over the world; they are made possible by the development of sets of facial muscles found in no other primates and apparently serving no other purpose than signalling emotions. The power to interpret them seems to be innate, since babies in their first weeks of life can already distinguish between a smiling expression and a grim one.

Our reactions to the attitudes of others has developed as dramatically as our power to interpret them. We can be seriously upset, not necessarily by the sight of bared fangs and the threat of physical attack, but simply by a coolness, a slight, a lack of respect. In one experiment, the subject was led to believe that he was playing an electronic game with several other people. He apparently enjoyed the experience until he was given the impression that the other (non-existent) players were increasingly, for no obvious reason, leaving him out of the loop. He asked himself the inevitable question: "What's wrong?" He felt rejected, and it hurt. Since the experimenter was making use of magnetic resonance imaging, nobody had to take his word for it that it hurt, nor even deduce it from his expression. A pain centre in his brain lit up as brightly as if he had received a slap on the face.

That pain centre is part of the human inheritance. We all recognise how it operates and how it stings, whether the occasion is rejection by one person whose opinion we value, or ostracism by a group we wish to belong to. If we speak in terms of purposive genes, the gene has planted it in the human psyche in order to influence our behaviour in a desired direction. The message it sends is: "The most important factor in your environment is other people. Do whatever it takes to get on with them or else find or found another group that you can get on with."

John Donne put it into five words. "No man is an island." We can, of course, respond to this injunction by telling the gene to go jump in the lake and retire to a hermitage, just as we can fight off other genetic instructions about eating when hungry and sleeping when tired, but we cannot deny that it is there.

It is not without precedent: few things in the evolutionary story are. A baboon, separated from its band, has been observed to hang around on the fringes of another group, enduring a series of rebuffs and tentatively working his passage into acceptance by menial tasks like babysitting for the females. But in humans, the desire for approval by others, like the sophisticated capacity to detect and empathise with their feelings, has expanded as notably as the capacity to think and communicate. Is this progress? Does it mean we have gradually become nicer as well as smarter? In more acceptable terms, has there been a process of directional variability proceeding in one direction for a very long time? The answer would appear to be yes.

Why has it happened? There were several factors that would have favoured it. One has been mentioned: the prolonged helplessness of human newborns. On the male side, there was an equally powerful incentive. We have descended from a fairly defenceless anthropoid species without sharp claws or fearsome canine teeth; their scuffles with one another could rarely have caused serious damage. But over a comparatively short period, as evolution measures time, these animals were transformed by the invention of weapons more lethal than the claws of the big cats or the fangs of the sabre-toothed tiger. It became a matter of life and death to be constantly alert to the moods of others and not to be entirely comfortable in their company, unless they seemed relaxed and contented.

Also, co-operation between humans pays off more handsomely than in any other species. There are hundreds of useful things that can be done by two pairs of hands which cannot be done by one; starting with something as simple as hunters carrying a carcass: "I'll take the front legs and you take the back legs."

This is not the case with animals that have hooves instead of hands. It is theoretically the case with other anthropoids but they seem not to have exploited it. If you were the selfish gene, what general instruction would you have issued to this anomalous survival machine you found yourself living in? Might it not be "co-operate?"

Not all the innovations work smoothly and we will need to ask why. If I have so far drawn a somewhat one-sided picture, it is in the interests of redressing the balance. We have lately become conditioned to believe that the most insuperable barrier to arriving at the truth about ourselves is human arrogance. We are warned against a subconscious desire to reinforce our sense of privilege and uniqueness by seeing ourselves as the pinnacle of the tree of life.

That was once a danger, but it is possible to become paranoid in our anxiety to avoid it. "Unique," like "progress," is not a dirty word. It is unrealistic to deny that some of the faculties we have acquired are as unique as the elephant's trunk. It is not necessary to believe that our uniqueness was predestined, or that it represents what all life was striving to become. It is not necessary to believe that we were pulled up to our biological pre-eminence by what Dan Dennett calls a sky-hook. It was due to natural causes. In Gouldian terms, one of those contributory causes must have been contingency. We struck lucky.

But it has happened. Agreed, serious scientists should not be thumping their chests and proclaiming how wonderful we are. But equally, they should not be hanging their heads and screwing their toes into the sand and muttering, "Aw, shucks, we ain't so special." Anyone is entitled to give vent to the value judgement of people like Ghiselin, who are convinced that human beings are the pits. But when they imply that this is a scientifically established fact, rather than a spasm of self-disgust, that claim should not be allowed to pass unchallenged.

> **Summary** *The evolutionary history of* Homo sapiens *has been characterised by an increase in the power of reason to prioritise between conflicting emotions. This has been accompanied by an increase in the variety and power of inherited eusocial instincts.*

You have damaged his self-esteem.

CHAPTER 18

WHAT'S LEFT?

So why is left-wing politics so contemptuous of biological theories of behaviour, leaving the Right to claim Darwin as its own?

— From the Evolutionary Psychology Website

That's an intriguing question, based on some very shaky premises. It suggests that anyone who criticises Evolutionary Psychology is automatically rejecting all biological theories, back to and including *The Origin of Species*. It confirms that the Right is now trying to claim Darwin as its own, but implies that this role was forced on it because everyone else had deserted him.

The aim of the new discipline, according to a manifesto in an early edition of the journal *Evolutionary Psychology*, is to "unmask the universal hypocrisies of our own species, peering behind self-serving notions about our moral and social values to reveal the darker side of human nature." The message here is that anyone exploring the positive aspects of human nature is (consciously or unconsciously) wearing a mask, craftily constructed to deceive the vain and gullible. Meanwhile the right-wing scenario represents the brave and naked truth.

If this formula is repeated often enough few people will be brazen enough to admit to a disinterested concern for others for fear of being branded as a prig, as being in denial and wearing a mask. That's a serious charge and it takes a thick skin to risk incurring it when most of your intellectual peers are boisterously defending themselves against it. "What, public spirited—moi?

Hey, listen guys, I'm just on the make like everybody else." Fair enough. The fashion in masks changes and there's a lot to be said for conforming if you hanker for a quiet life. But if anyone claims that this is not merely a change of attitude but a new scientific truth, they need better evidence for it than they have yet adduced. And they should not be too surprised if non-scientists who cannot follow the arguments mutter to themselves: "That is not how it feels," and scuttle into the arms of the Creationists.

It is not as easy as it used to be to define what is meant by left-wing. Throughout the 20th century, the term was closely associated with the teachings of Karl Marx. His economic researches convinced him that it should be possible to construct a scientific Socialism that would predict the future course of economic changes. In the minds of many, his predictions of the imminent collapse of the capitalist system were converted into a dogma which filled them with hope and confidence and solidarity. But Stalin's exploitation of that faith turned the Soviet Union into a totalitarian nightmare before it finally fell apart, and left its erstwhile defenders feeling disillusioned and rudderless. So what remains of left-wing beliefs and attitudes since the Marxist prophecies were discredited?

What remains is something that was there before Marx and before Darwin. In every complex human society, there is a spectrum of temperaments, or personality traits, or ways of looking at life, in respect of degrees of empathy, optimism, acquisitiveness, trust, sense of fairness, fear of change, and other such characteristics. Some of these features tend to clump together, rather as atoms clump together in molecules, to form the opposing personality profiles recognised as right- and left-wing attitudes. Like other human characteristics they are determined by the interaction of nature and nurture. At different times and places, different proportions of a given society will accumulate at one end or another of the spectrum.

In modern societies, people at one end are more responsive to the ideas of left-wing philosophers and economists, those at the other end to right-wing ones. From time to time the Right or the Left may be disconcerted by the failure of Marx's prophesies or the validation of Galileo's, but the underlying spectrum persists. It is a fantasy to imagine that one day it will disappear because everyone will see the light and migrate permanently to one pole or the other. It is a commoner fallacy to believe that all people at one end are clear-sighted and truthful, and those at the other deluded and deceitful.

At present the darker view of humankind and its prospects is the one being most vigorously promoted. To the true pessimists, *Homo sapiens* at his best was never much to write home about, and already the evolutionary rot is setting in. They point out that our brains, after a growth spurt which expanded them by at least one third in less than a million years, have now begun to shrink. Modern skulls are not only smaller than those of the Neanderthals, but 6% smaller than in early modern *Homo sapiens* one hundred thousand years ago. We really don't need to lose any sleep over that. There were doubtless good reasons, probably obstetric, for halting the expansion of the cranium, but Mother Nature is as adept at miniaturising her inventions as any chip manufacturer in Silicon Valley. The bat's minuscule sonar equipment works at least as well as the dolphin's.

The reassuring fact is that none of us uses more than a fraction of the cognitive capacity we already possess. As with the capacity to speak, the capacity to reason depends on the social environment if it is to be fully developed and expressed. But if we view it in terms of achieved intelligence rather than mere innate potential, the evidence still indicates that our species is certainly not dumbing down. On the contrary there is a worldwide phenomenon known among IQ investigators as the Flynn Effect, substantiated in many countries and on many different types of IQ tests. By all available testing procedures, the national average score since World War II has been moving upwards by as much as a point a year. *The Bell Curve* confirms this and concedes that it is not merely a side effect of more widespread literacy, since most of the change has been concentrated in the non-verbal portions of the test. There are all kinds of scarce resources that might hamper our chances of improving the human condition, but lack of the native ability to think is not one of them.

Indeed, there is one encouraging aspect of it which has only recently been fully appreciated. The power to reason may have originally emerged to serve the organism by monitoring its primal appetites and increasing the chances of their being fulfilled. But at some point, it became a player as well as a referee. The earlier claim I made that "reason cannot tell you what to want" was not entirely true. Problem solving was sufficiently adaptive to be endowed with its own emotional behavioural reward, just like other basic appetites.

We have been accustomed to read about the "Aha" experience, the Eureka moment of discovery that allegedly projected Archimedes out of his bath to run naked through the streets. It was an extreme illustration of an experience familiar to all of us. Things we cannot explain make us feel uncomfortable. When we find a solution to puzzling questions, when the penny drops, the sensation is enjoyable. These sensations are caused by the activation of pain and pleasure centres in the brain, as specific as those that govern bodily appetites, and the magnetic resonance imaging screen can pinpoint the specific areas that light up. By some people that sensation is so highly valued that they conduct a restless search for problems in need of solving. It can become a mild addiction, catered for by newspapers that print crossword puzzles. For some people it becomes as compulsive as any other addiction, giving rise to the old definition of an intellectual as a person who has found something more interesting than sex.

Not everyone finds this cerebral capacity reassuring. Equally part of human nature are the passions, some of them destructive, and strong enough to hijack reason for their own ends. When that happens, improved technological ingenuity only ensures that the kind of conflict which once killed hundreds of people may now kill millions. What we need is to become not cleverer creatures but better ones. Unfortunately, no one has been able to invent a way of measuring good and evil because, in the matter of ethics, one man's meat is another man's poison. One man's courageous freedom fighter is another's evil terrorist.

Scientists avoid speaking of good and evil because of this subjectiveness, and because they see themselves as professionally debarred from making value judgements. Yet these concepts are powerful determinants of human behaviour; so powerful that some people will go to the stake for them. They must have come from somewhere and no Darwinian account of the human organism can be complete without them.

One possible definition is that good and evil are whatever a given society of human beings communally approves or disapproves. We are not now talking about genes but about John and Thomas and Mary and Elizabeth. Characters like these have been around from the beginning, even if we stick silly names on them like Ig and Og. They have been living together with others of their own kind, wanting to be accepted, fearful of being betrayed, gossiping about their own experiences and about others' behaviour and ceaselessly trying to edge towards a consensus about how people should expect other people to behave. Knowing what to expect of one another makes it much easier to live together.

Robin Dunbar was the first scientist to apply the layman's word "gossip" to this process and to suggest it might have been a major driving force behind the emergence of language. For a long time, early humans seemed to have been thin on the ground and living in fairly small, self-contained groups with a way of life that remained unchanged for long periods. These conditions would favour the emergence of generally acceptable rules of conduct— of "right" and "wrong"—which would serve to minimise social friction. Behavioural norms could differ between different groups, rather as they came to differ between the rhesus monkeys and the stump-tailed ones. Up to a point, the fact that they were agreed on was more important than what they were. As with the rules of the road, it doesn't matter whether you drive on the right or the left, as long as you all agree on which is correct. But it was very likely that some minimal basic kit of rules for social harmony would come to pertain to the whole species—and there is some evidence that this occurred.

In the 1950s, Erving Goffman went to stay on one of the smaller islands in the Shetlands. It was a place where people knew their neighbours, and their way of life did not change very much. Goffman moved in there and watched the people and listened to them and made notes, with the detailed attention that

Goodall bestowed on her chimpanzees. He observed that a good deal of what they did and said, in the course of a day, was in the nature of mutual reassurance—a confirmation that social relationships were in good working order and functioning smoothly. For example, on any given day, if someone passed in the lane an acquaintance that he saw every day, the two would exchange a brief greeting. If one had been away for a month, something more was called for: an exchange of questions and answers would mark his return and reintegration into the local social fabric.

We are all subconsciously aware of these rules. We know if we meet the same person again a few hours later, no words are required, just a flash of eye contact and recognition. But even then, if that courtesy is omitted and he "looks through you," you may wonder what's eating him. Goffman could pick the bones out of what is said when someone asks the time of a stranger without a watch, and detect what percentage of the words spoken constitutes an assurance of basic goodwill, indicating an unspoken social contract, "I won't try to make you feel bad if you don't try to make me feel bad."

Once he had published his observations, anthropologists looked for parallels and found some such code operating in communities all over the world—sometimes vestigial and sometimes highly elaborate but always there. It is vestigial, for example, on the pavement of a metropolis where people are surrounded chiefly by strangers. The only etiquette there is to mutter an apology if you barge into somebody and, as in a crowded lift, not to stare. But in cities too, there are subsets of acquaintances who constitute mini-communities, and it's unusual to glimpse someone you know in the crowd without a signal of recognition and a "Hi" or an "Awright, mate?"

These things are as old as the Pleistocene and older. We can be confident of that because a version of them is seen in many social animals. A young wolf or chimp that has been missing for a few days gets, on its return, twice the greeting and nuzzling and tumbling over that it would receive if it had never been away. But in our own case, communities did not remain small and mobile: they became more static and larger. They stratified into castes and hierarchies, and that changed the rules. If under some misapprehension you ask the time of someone who considers himself your social superior, you could find that mutual reassurance is not on the menu. Simply by making the request, you have damaged his self-esteem. An MRI machine would probably detect a sensitive pain area lighting up in his brain, caused by your failure to acknowledge what he hoped would be blindingly obvious to anybody, namely that he has blue blood, or is very rich, or belongs to a nobler race. In these conditions, what he expects from you is not mutual respect but deference.

It is the stratifying that complicates the rules of how we should behave. The communally constructed rules of how to get on with one another are flexible and can be revised to reflect changing circumstances, but in hierarchical societies, they are overruled by instructions coming down from above. There are decrees issued by chieftains and kings to be obeyed on pain of punishment, and there are commandments from the gods, interpreted by medicine men, oracles, sages, priests and bishops, to be obeyed on pain of divine vengeance. These are proclaimed to be immutable, and to flout them is not just bad manners to be frowned on. It is treason; it is heresy; it is evil.

And that is where Right and Left come in. Very roughly, right-leaning people hold that hierarchies are at the heart of civilisation. They preserve order, they ensure continuity; only wicked men would want to weaken them. Left-leaning people don't imagine, as once they did, that hierarchies can be dispensed with altogether. But they have a stronger sense that the horizontal divisions are the more superficial ones and are designed for the convenience of the minority. That is a quite widely held view, and has been memorably expressed many times from "When Adam delved and Eve span, who was then the gentleman?" to "The rank is but the guinea stamp, a man's a man for a' that," and the words that every American child is taught to honour, "We hold these truths to be self-evident: that all men are created equal."

It is not a newfangled idea: it was going strong in the Pleistocene, though it is not an aspect of that era that is often stressed. Anthropologists have studied scores of contemporary hunter/gathering societies, spread over four continents, and concluded that they are characterised by "egalitarianism, co-operation and sharing on a scale unprecedented in evolution. In fact, rank is simply not discernible among hunter/gatherers. This is a cross-cultural universal which rings out unmistakably from the ethnographic literature sometimes in the strongest terms."

That should not be confused with any wishful philosophising dreams about the noble savage. It is not based on idealism, but on observation of the lifestyle which early humans shared for so long, and which is so often acclaimed as a determinant of our present behaviour. It does not pretend they were any milder or kindlier or wiser than we are—but they were more egalitarian.

Most of us nowadays live in societies where that degree of equality is unsustainable, and being an adaptable species we have learned to live without it. When stratified systems like feudalism last long enough, they are ultimately taken for granted even by those on the lower rungs, as long as they know where they stand, are accepted and respected by others of the same status, and can take a pride in the work they do. The traditional rights asserted by their overlords became partly mitigated over time by the acceptance of a number of traditional obligations, if only those popularly known as "bread and circuses." The women among the gentry send soup to the ailing poor and the lord provides wassail at Yuletide. But the egalitarian system lasted longer, and has deeper roots.

Richard Wilkinson, in his booklet *"Mind the Gap,"* assembled a good deal of evidence that humans do not function best in societies with a high degree of inequality, and he described the psychosocial costs incurred. He demonstrated that the striking differences between health and life expectancy in different levels of society are not exclusively due to material factors such as food and housing and pollution and the physical demands of labour. "Research findings have gradually forced us to see that what matters most is psychosocial welfare and the quality of the social environment." Among the developed countries, it is the most egalitarian that have the highest life expectancy—not the richest. Greeks have less than half the average income of Americans and yet are healthier. In the USA, a strong relationship between income inequality and death rates has been reported for the fifty states and the 282 standard metropolitan areas. This effect is independent of average living standards, the proportion of the population living in absolute poverty, expenditure on medical care and the prevalence of smoking. Absolute deprivation has long been recognised as harmful to health, but Wilkinson's point is that "the indirect, psychosocial effects of *relative* deprivation are unexpectedly powerful."

Another striking example was the United Kingdom during the two World Wars when income differences narrowed dramatically. On both occasions there was a palpable sense of camaraderie in the civilian population, which those who lived through the second conflict can vividly recall. Food and clothing and petrol, all in short supply, were strictly rationed. Anyone who had enough money or influence to get more than their fair share was anxious to hide the fact rather than flaunting it. Conspicuous consumption was considered treachery to the nation and the war effort. In those circumstances, even though the national resources were in desperately short supply, civilian health improved two or three times faster than normal.

It would be hard to prove that the feeling for equality is, in the technical sense, innate. But there are a few pointers suggesting that it might be. It is particularly strong in children. Once they have become conscious of the words that reflect their society's moral values, they are very quick to commandeer the ones they approve of, even while their general vocabulary is still very limited. It must be admitted that one of these examples seems to support property rights: "That's mine!" But just as quickly they grasp the egalitarian ethic, and are genuinely outraged when it is not adhered to. "That's not fair!"

It is relevant too, that the feeling against inequality is not limited to those on the losing end. People, not all of them but a good percentage of them, feel uncomfortable in the presence of others much less well off than themselves. The discomfort can be alleviated by keeping at a distance from them, but it is harder to avoid knowing that they are there. Anthony Trollope, in his autobiography, described the sensation. "We who have been born to the superior condition—for in this matter I consider myself to be standing on a platform with dukes and princes and all others to whom plenty and education and liberty have been given—cannot, I think, look upon the inane, unintellectual and toil-bound life of those who cannot even feed themselves sufficiently by their sweat, without some feeling of injustice, some sting of pain."

If you adhere to the conventional wisdom, you can rationalise away this sting of pain; you may say it is simply an adaptive reaction based on the fear of envy. Living with envy, even when it is impotent, is not pleasant. Shakespeare's Macbeth described how the joys of power can be curdled by losing the ability to trust anybody, leaving only the feeling of being surrounded by "Curses, not loud but deep, mouth-honour, breath, Which the poor heart would fain deny, and dare not."

Tyrants do not rise to power only in hierarchical societies based on the supposed superiority of a single race or class. They have also risen to the top in societies which were founded in the name of equality, like the Soviet Union, just as greed and oppression have repeatedly risen to the top in churches founded in the name of humility and love.

The Left today is in a state of disarray and disillusion, still trying to work out what went wrong. There are voices telling them that what went wrong was that their aim was always inherently impossible, that humankind never was created equal, and never evolved even to hanker for fairness. It evolved only to seek power and privilege, so that every one of us in our secret hearts is red in tooth and claw and crying, "Devil take the hindmost." It is a caricature of human nature which the Far Right has an interest in promoting. That is why it is so eager to spin-doctor Darwin's message in the hope of claiming him as its own.

> **Summary** *There is no basis for the suggestion that the aims of the Left are any less in tune with evolved human nature than those of the Right.*

...making love, not war.

Chapter 19

The Magic Hand

This book is about the moral, emotional, and political coloring of the concept of human nature in modern life.

— Steven Pinker

The book in question is *The Blank Slate*. It is lucid, authoritative, earnest, and compellingly argued. This has been recognised by many of Pinker's fellow Darwinists whose scientific judgement I deeply respect, and who cannot understand why anyone but a deranged left-wing extremist should take exception to anything in it.

But Pinker makes one further claim for it which I do dispute. He represents his approach as moderate, impartial, and coldly analytical, and specifically not advancing the agenda of the political Right or Left. He explains in the first five pages that he simply wishes to point out that the "so-called radicals" have now taken over the academic establishment, and are exploiting this position of power to promote ideas which put blinkers on research, disconnect intellectual life from common sense, flaunt fantastical beliefs, show contempt for truth, logic, and evidence, force millions of people to live in drab cement boxes, and demand the release of psychopaths who promptly murder innocent people. What could be more objective than that?

In Alper's phrase, there is plenty of politics in that book and I didn't put it there. A main thesis running throughout the book is a validation of the Tragic Vision. The term is taken from

Thomas Sowell's book, *A Conflict of Visions*, outlining two different ways of thinking about the human species in a political context. There is a Tragic Vision in which humans are inherently limited in knowledge, wisdom and virtue, and all social arrangements must acknowledge those limits. Opposed to it there is a Utopian Vision in which the human capacity to be cooperative is as great as its capacity to be competitive and if social circumstances can be improved, there is every hope that behaviour will improve accordingly.

Pinker explicitly endorses the Tragic Vision and the social arrangements which should flow from it. His repeated assurance of "really do...really does" is aimed at hardening it from one way at looking at life into the only possible way, stamped with the imprimatur of the scientific method. He is impatient with any suggestion that some "mysterious drive towards goodness" forms any part of human nature. He rebukes Stephen Jay Gould for claiming that *Homo sapiens* is not an evil or destructive species, and for suggesting that well-intentioned people outnumber other kinds by thousands to one. He cleverly catches him out on the statistics, pointing out that psychopathic males, who are not at all nice, make up as much as 3% or 4% of the male population (forget the females, they would only distort the picture) and not just a few hundredths of a percent, as Gould maintains.

Pinker can't help seeing that even three or four baddies out of a hundred is not very impressive, so he suddenly demands, "Also, does it make sense to judge our entire species as if we were standing *en masse* at the Pearly Gates?" Maybe not, but his whole book is devoted to *en masse* statements about our entire species. He cannot reasonably claim that Gould is judging while he himself is only describing, because when he describes our primal urges and the thought processes of his opponents, his vocabulary is as emotionally loaded as that of any writer who ever dealt with the subject, with the possible exception of Raymond Dart.

In short, Pinker's portrait of humanity closely resembles *Homo economicus*, the model currently favoured by academic economists. Their calculations are all based on the assumption that every man, woman and child is engaged in a zero sum poker game against every other man, woman and child, with the sole intention of coming out ahead of the game. Just occasionally the theorists are assailed by lingering doubts about this assumption; that is why they have taken such a keen interest in recent trends in evolutionary theory, hoping to have their doubts resolved.

There was, for instance, the curious case of the market for blood in the middle of the last century. The demand for blood was rising rapidly and blood banks in the United States decided to boost the supply by paying donors. In Kansas City a community blood bank, organised by volunteers, was condemned as a conspiracy to restrain commerce and closed down. In Britain, the converse applied; it was the *sale* of blood which was prohibited, not from any Puritanical dislike of the profit motive but because people desperate enough to sell their blood for cash were thought rather more likely to suffer from communicable diseases. The National Health Service promoted a scheme whereby blood donors were rewarded with a "thank you very much," a cup of tea and a biscuit. Between 1956 and 1957, the blood supply in England and Wales increased by 77% and in America, it increased by 8%.

This can be worrying for economists who need to feel they have a finger on the pulse of the consumer and know what makes him tick. So what's this with the cup of tea? The same puzzlement has been illustrated by quoting a famous Monty Python sketch where a merchant banker is being asked for a donation.

> Banker: I'm awfully sorry, I don't understand. Can you explain exactly what you want?
> Ford: Well, I want you to give me a pound and then I go away and give it to the orphans.
> Banker: Yes. . .?
> Ford: Well, that's it.
> Banker: I don't follow this at all. I mean, I don't wish to seem stupid, but it looks to me as though I'm down a pound on the whole deal.

They are made slightly uneasy by the old song of the village cobbler who knew that if he did a really good job, his customer would not return for a very long time. "The better my work, the less my pay, But work can only be done one way." Far more alarming was the short period in the sixties when marketeers felt the ground beneath their feet begin to crumble, as thousands of young people, many from prosperous homes, declared that the rat race and conspicuous consumption were yesterday's fads, and took off for San Francisco and the simple life and making love not war. If that attitude had seriously caught on, the mainspring of the system might have snapped.

There have been slight after-shocks of that experience in the last few years—nothing flamboyant, but the number of people quietly deciding to "downshift" out of the rat-race is climbing. In Britain, more than three million chose to look for jobs with less money and less hassle in 2004, a rise of 200,000 in a twelve-month. In Europe the figures are up from 9.3 to 12 million since

1997. It raises the question economists dread: "What do they want if they don't want money?"

In the drive to win hearts and minds for Conservatism, the Tragic Vision appears to beef up the intellectual content of right-wing propaganda, but it does little for the 'hearts' end of the formula. However Pinker has found an answer to that; he wants to cross-fertilise the Sowell thesis with the Expanding Circle thesis of Peter Singer, and produce a hybrid that will be acceptable to all.

Peter Singer believes that human beings have an in-built tendency which causes them to regard other people as targets of sympathy, and inhibits us from harming them, and whereas it was once extended only to a small section of the human race, it has spread steadily outwards to embrace the tribe, the clan, the nation, the race and in some degree to all humankind, including women and children and criminals and the disabled—and is now beginning to extend to other animals. I find this proposition entirely convincing. I would imagine many people on the Left feel the same way. More surprisingly, so does Steven Pinker. Here is his credo: "For all its selfishness, the human mind is equipped with a moral sense whose circle of application has expanded steadily and might continue to expand as more of the world becomes independent."

This doesn't fit in too well with the image of *Homo economicus* who is not supposed to go around brandishing a moral sense. Pinker hastens to assure us that when he speaks of faculties underlying empathy, foresight and self-respect, he is not going mystical on us. These qualities may not be primal urges but they are something more than memes; they are "physical circuits residing in the pre-frontal cortex and other parts of the brain, not occult powers of a poltergeist." (All those who had been stupidly insisting that they were occult powers of a poltergeist must now retire abashed). He even speaks of a "knob of sympathy" that may be in there somewhere. "It could have arisen from a moral gadget containing a single knob or slider that adjusts the size of the circle embracing the entities whose interest we treat as comparable to our own." Moreover, it is here to stay. "Long ago, these endowments put our species on a moral escalator." The escalator has carried us up to this point and may be expected to carry us further still.

This is amazing stuff; it reads like Professor Hayek on the road to Damascus. There are only three aspects of it that need examining a little more closely. One is his account of the escalator, offering a vision of the whole of society being wafted up effortlessly and unanimously to a higher moral level. Another is his prediction of what the next steps upward might consist of. And the third is the question of what he means by "as more of the world becomes independent." Independent of what exactly?

First, the escalator. He expresses his retrospective approval of the advances that have already been made; it was obviously right to grant more civil rights to women and Afro-Americans. He has erased from his memory any recollection that when these proposals were first mooted, it was the Left that campaigned fiercely in favour of translating them into legislation and standard practise, and it was the Right which resisted bitterly every step of the way. In the sixties, those demanding reform were not welcomed as fellow passengers on the moral escalator; they were reviled as agitators or diagnosed as psycho-pathologically disturbed.

One philosopher, Professor Howard S. Schwartz, wrote a weighty volume concluding that political correctness is a way of defending ourselves against our own self-hatred. It was "an attempt in the name of the primordial mother to expel the father and the external world he represents and to substitute the unconditional love of the mother." Perhaps if he could have put them all on the couch, he could have cured them. A symptom of their malaise, he states, was a belief in "the legitimisation of coercion," betrayed when students resorted to strong-arm tactics like sit-ins. These methods were indeed employed in some instances and were discouraged by calling in the troops. That was one way of convincing them that coercion could never be legitimised.

The tactics of the Left in the sixties did tend to be rowdy. As with any shift in the balance of any power structure, there were a few individuals among the beneficiaries of change who grew shrill and spiteful, and there were good and gentle people who got hurt. They were good people in the ways that all their friends had always regarded as good. Some were astonished to be told for the first time that women and blacks would all along have liked to be treated as equal to men and whites. The possibility had simply never crossed their minds, yet now in some instances it was threatening their careers. They didn't deserve to be hated—but it is hard to know how you make somebody realise you don't like what they are doing unless you say so loud and clear, and if they seem not to be listening, shout. In fact the bloodless tactics they used, as in the Martin Luther King marches and the Women's Lib demos, have been acclaimed in other contexts as those of the "velvet revolution."

Regime change can never be pleasant for everybody. Schwartz, writing of the events in academia, recalled at one point with a note of horror "it had become about who was good." A new flag had been planted on the moral high ground and the people who had always run the place had no suitable weapons to use against it. They wanted to say something like, "This place is concerned with the intellect. A university is not about being good." The implicit answer was: "It is now."

But the sixties and seventies are fading from the memory and the Left has lost its exuberance and its certainties. The Marxian prophecies were not borne out. Capitalism did not collapse under the weight of its internal contradictions: it was Soviet Russia that collapsed. In socialist totalitarian countries, as inevitably as in capitalist totalitarian ones, absolute power corrupted absolutely. People with left-wing convictions no longer quote the phrases of Marxist dogma in the way that fundamentalists quote texts from the Bible—and for that we must be thankful. Dogma is a useful weapon for ensuring that everyone marches in step, but it is always a hindrance to clear thinking.

The tools of action as well as the doctrines need to be adapted to changing circumstances. At one time, the power of organised labour represented a serious counterbalance to the power of big business. There were trade unions of skilled workers who were indispensable because it would have taken years to train replacements for them, but manufacturing processes are continually broken down into small, disparate procedures that can be carried out by clever machines whose operators can be readily replaced. New industries like IT have never depended on a local workforce with a strong sense of solidarity. If you tried to stage a strike in a Western call centre, it would only speed the day when it would be closed down and the jobs transferred at a stroke to a cheaper labour market in the East.

When people get sufficiently disheartened, even democracy becomes less of a safeguard. People begin to suspect that the media and the politicians are for sale to the highest bidder. They no longer know what to believe, and the poorest sections of the community are the least likely to use their votes. However, there were people whose hearts were on the Left before Marxism and before Trade Unionism and they are still there, and when it is time for the next step up the moral escalator their voices will be heard.

One of the issues that can still bring hundreds of thousands of them out into the streets is the issue of war. The attitude to this has been changing very slowly over the centuries; at one time it was full of glamour, the only trade fit for a gentleman. In England in 1914, young men were straining at the leash for the opportunity of fighting for their country. "Now God be thanked, who has matched us with His hour."

Killing was good. In the Bible, we read that Saul hath killed his thousands but David hath killed his tens of thousands—even better. When Alexander led his armies halfway across Asia, he did not need to persuade his followers that the people they slew were evil, or a danger to their own people back in Macedonia, or to humanity. They were there to be conquered, and conquering was what he enjoyed doing.

If attitudes had not changed at all since then, Harry Truman would have commissioned a huge, triumphal monument the size of a pyramid boasting of the holocaust of nearly two hundred thousand Japanese in three days, a world record. But that episode is not recalled with much exaltation. The concern is to justify it in terms of necessity and self-defence, as the lesser of two evils. Now that war is being recorded by cameras instead of ballad-mongers, it seems uglier and more intolerable. For wars involving the great powers we still put out the flags and bang the drums, but in more local conflicts, a note of exasperation creeps into the comments of non-combatants. "What's the matter with these people? Are they stupid? Why can't they see that there must be a better way of solving these problems? Why can't they just sit down together and embark on a peace process?"

Pinker does not envisage a revulsion against war as the next stage in society's moral advance. Indeed, he has some difficulty in imagining any way in which society can possibly improve. Given his view of what people are made of, he tends to think that maybe this is as good as it gets, and expresses the hope that nobody will come along and spoil it. However, having hitched his wagon to the concept of the expanding circle, he feels called on to give some examples of the way in which we might continue to move onward and upward, and he mentions two possibilities. He suggests that people might expand the circle of their social concern to embrace an empathy with zygotes, blastocysts and fetuses, and with the brain-dead, and with the ecosystem. The ecosystem sounds fine in the abstract, but he does not go into detail about it. I suspect that any concrete plans, such as proposals to ration the consumption of fossil fuels, might strike him as a loony Left plot in restraint of trade. So we are left to anticipate the next step forward as a ground-swell of popular support for two specific policies, the opposition to contraception and the opposition to voluntary euthanasia.

There remains the question of what he meant by saying that the expanding circle might continue to expand as more of the world becomes independent. Independent is a beautiful word but an ambiguous one. For the founding fathers of his country, it meant national independence from the rule of foreign imperialists, and I don't think he can have meant that. There isn't much of the British Empire left outside of the Falklands and Gibraltar, and they both seem quite anxious to stay in. The Russian hegemony has broken up, and the United States has never sent out waves of emigrants to settle in foreign lands because they are better off where they are. The people still clamouring for that kind of independence are small groups like the Chechnians and the Palestinians and the Basques. I can't see Pinker hailing them as freedom fighters. The Taiwanese used to be on America's list of places deserving liberation from a foreign yoke, but since China opened up its markets, they too have been struck off the list and told to calm down.

Perhaps "markets" is the clue to the mystery. Independence to Pinker means freedom from interference with market forces. The French have a word for it, which is *laissez-faire*. There is no need to deduce that this is his meaning because he explicitly spells it out. "The Tragic Vision looks to systems that produce desirable outcomes, even when no member of the system is particularly wise or virtuous. . .Market economics, in this vision, accomplish that goal." Consumers can, he believes, react by following a few simple rules and the invisible hand will do the rest. He implies that this process is not only compatible with the expanding circle of empathy but is the prime mover of it. He points out that the circle has expanded in tandem with our physical circle of "allies and trading partners," for the obvious reason that "you can't kill someone and trade with them too." I hope he did not mean that too literally. It would imply that if any country declines to trade with yours, on your terms, you are morally entitled to go in there and kill them.

I don't for a moment suggest that Pinker is unique or even unusual in attempting this fusion of fundamentalist economics with Tragic Vision biology. His economic creed is just a somewhat simplistic version of the beliefs currently in vogue among his academic peers and contemporaries. Let's fill in a bit of the background.

The founding father of economics was Adam Smith whose book, *The Wealth of Nations*, was published in 1776. He used a simple scenario based on small-scale, everyday transactions to illustrate his idea and introduced us to the archetypal butcher, who does not supply us with meat out of benevolence but out of regard to his own self-interest.

Smith concluded that in such transactions, the selfish drives of many individuals are transmuted as by an invisible magic hand into the good of all. It's a neat idea and it works. Peoples' needs are constantly changing, the availability of resources is constantly changing. In many transactions, the profit motive is the only tool flexible enough to relate the supply of goods and services to the demand for them in our complex modern world.

Soviet Russia tried to replace the reliance on greed and profit with a totally managed economy, where central planners would divine in advance what people would want (or what they ought to want) and that system did not work. But a totally uncontrolled economy does not work either. It works on paper, on the assumption that everyone's decisions on what goods to buy or what wages to pay are purely logical. It also presupposes that all the buyers and sellers, all the employers and employees, have a complete knowledge of all the options open to them and operate on a level playing field where everyone is free to accept or decline what is on offer. The golden keyword is Choice.

One snag with choice is that beggars can't be choosers. Hunger is not rational; if you have no money and no food, you cannot say, "The wages you propose are unacceptable, I'll wait for a better offer." Another snag is that a firm driven by the profit motive will try to buy up all its competitors and form a monopoly. It will freely choose to destroy its competitors' power of choice. The bigger the imbalance of power between the rich and the poor, the smaller is the chance that the magic hand will work for the good of all. When a farmer negotiates with a supermarket, he ends up by getting ten cents out of the $3.99 selling price for a box of cornflakes.

The fact is that you cannot extrapolate the rules of the village baker and candlestick maker to the conglomerates and multinationals and to corporate entities like Enron and Worldcom. In Adam Smith's concept, the entrepreneurs were called things like Williams & Son, Ltd. They made end-products like cloth and steel and pottery, and wanted their companies to survive them and pass on to their descendants, so they tried to build a reputation for square dealing. In the world of high finance, the biggest profits are not made by producing anything, but by buying up enterprises that somebody else has created and after a bit of asset-stripping moving on and out.

It is not merely the underprivileged who are operating without a knowledge of all the facts. The money market deals in the future, and about the future we can only guess and gamble. The Great Depression that followed the Wall Street crash in 1929 was not caused by any drastic change in the real world of commodities and service. It was due to the pipe dreams and hyped-up expectations of speculators obeying the promptings of greed, as the theorists recommend us all to do if we hope to be saved. The results were disastrous, so for a period after that, there were politicians like Franklin Roosevelt taking advice from economists like

J. M. Keynes. They worked on plans for a mixed economy to protect society against the excessive rapacity that could inflict such widespread damage on the lives of millions of people.

Consequently, the degree of inequality between the rich and poor began to diminish. Sir Isaiah Berlin was expressing a commonly held opinion when he said, "The liberty of the strong, whether the strength is physical or economic, must be restrained." But since then, starting with Milton Friedman and continuing through teachings like that of Margaret Thatcher's favourite guru, F. A. Hayek, to the current neo-Conservative doctrine of Public Choice Theory, economists have drifted back to the old romantic idea of the magic hand. They hold that any tampering with the free market by any Government is doomed to failure and will do more harm than good. Their doctrine has been summed up as "those who posit a collective good or an ethic of public mindedness are mere sentimentalists pursuing an unscientific mirage."

Nobel prizes are awarded to economists like Kenneth Arrow and James Buchanan who favour this thesis. Governments are seen as a potential danger to the system, though somewhat less dangerous than they used to be because, as winning an election becomes ever more costly, those who contribute most to the coffers gain more and more influence over the manifestos.

The ultimate dream is strangely enough one that the Marxists used to share, the dream of the withering away of the State. A recent book entitled *The Company* proclaims that the company, or the corporation, is "the most important organisation in the world, the basis of the prosperity of the West and the best hope for the future of the rest of the world." The dogmatic certainty that was once typical of Marxism now attaches to the devotees of global capitalism. The results of this are clear: within States and between States, inequalities are rapidly widening. Robert Kuttner sums up the position in the United States:

"The widening of inequality, beginning in the mid seventies and accelerating in the eighties, is one of the best documented recent economic trends. No matter how you measure it, the income distribution in the United States has become more extreme. In the period between 1979 and 1993, the top 20% gained 18% while the bottom 60% actually lost real income, and the poorest 20% lost the most income of all, an average of 15% of an already inadequate wage. . .All of the gains to equality of the post-war boom have been wiped out."

In Britain, too, while the overall wealth of the nation has increased, the gap between rich and poor, even under a Labour Government, continues to widen. The government's economic policies are guided by the Adam Smith Institute, an ultra-right-wing lobby group which had advised the previous conservative government to sell off the railways, deregulate the buses, cut top rates of income tax, introduce the poll tax, and start to part-privatise the health service and education. Last year the Institute received 7.6 million pounds from the Labour government for continuing to offer the same line of counseling, and in fact more than half of its income is now supplied by the British taxpayer. Under its influence, many people have become considerably better off, while the poorest are stranded in pockets of intractable deprivation. Their plight is becoming institutionalised and rechristened as the phenomenon of "social exclusion."

The gap between rich and poor countries tells the same story, even more strikingly. In the name of the new doctrines, millions of pounds of "Aid" are poured into the hands of crony-capitalist rulers like Suharto in Indonesia and Imelda Marcos in the Philippines. Within such Third World countries, as well as between them and the First World, the gap continues to widen. According to a UNESCO report in 2002, half of the world's population subsists on less than two dollars a day. In 2003, the citizens of the United States consumed on average 88 times as much of the world's energy resources as the citizens of Bangladesh. If these trends were to continue, we would end up with half the world suffering from the effects of malnutrition and the other half from the effects of obesity. The next step up the moral escalator must surely be the recognition that this cannot be a good idea. As Peter Singer concluded, if we say that this is just the way the world is and always will be and there is nothing we can do about it, we are not part of the Left.

One thing that could get in the way of doing anything about it is the claim that science is bestowing its seal of approval on the basic vision promoted by neo-con economists. Their faith in that vision is so strong that they feel qualified to make prophesies (as Marx did) about the course of future events.

It is time to take a look into their crystal ball.

> **Summary** *Pinker's Tragic Vision of human nature is a depressing one. But he believes that, by an extraordinary stroke of luck, the laws of supply and demand have the power to exorcise all possible ill-effects of human weakness, so that if they are not interfered with everything will come up roses.*

Pity about the moa.

Chapter 20

The End of History?

I know it is at present the received opinion that the competitive or "Devil take the hindmost" system is the last system of economy that the world will see; that it is perfection, and therefore finality has been reached in it...

— William Morris

That was written in the 1880s. In the 1990s an even more final finality had been arrived at, and it is often summed up in a phrase used by Francis Fukuyama as the title of a book published in 1992: *The End of History*. It is the mirror image of the Marxist prophecy that once a truly socialist society had come into existence, its virtues would be so resplendent that all nations would yearn to emulate it and the world would enter into a promised land of prosperity and harmony.

The claims of the new economists harmonised remarkably well with the views of human nature being advanced by the new school of evolutionists. Steven Pinker was by no means the first to make the connection. Matt Ridley was writing in 1996: "I am not going to fall into the trap of pretending that our dim and misty understanding of the human social instinct can be instantly translated into a political philosophy." Such a preamble invariably expects to be followed by a "Nevertheless..."—and it was not long in coming.

One of the main lessons he thinks politicians can learn from biologists is that governments are a very bad thing. His reasons for saying this are simple. He describes how various small groups of lobster-fishermen, nomadic herdsmen, medieval wood-cutters and hunter-gatherers on the Pleistocene model have traditionally managed their affairs. These are conducted, he points out, in sensible, virtuous, sustainable ways by local people. The resulting regimes are not only egalitarian but also green: they do not damage the environment or deplete natural resources. (Pity about the moa, but then nobody's perfect.)

This account is oddly reminiscent of Rousseau's noble savage, although Ridley has insisted that only the soft Left believes in noble savages. He labels what his lobstermen are doing as private enterprise, which to him puts them into the same category as Enron and Wal-Mart and Standard Oil. Therefore he reasons that if any government should try to interfere with any of these operations it would be acting as the enemy of human happiness. It is sobering to read his analysis, because his heart is so clearly in the right place. He deplores the same things that I deplore, he blames them on government interference as confidently as I blame them on global capitalism, and it may well be the case that we are both oversimplifying. My point here, as throughout the book, is that there is more than one way of looking at things, and I am tired of being told that reason and Darwin and Science with a capital S have proved that only one of them is right.

The kind of capitalism now being commended is not the same model that was extolled by Adam Smith. Private enterprise was then the protégé of the state. The British government had prepared the way for it by the Enclosure Acts which drove much of the population off the land and into the mines and the mills, and by enabling legislation on the lines of Limited Liability Acts and patent rights.

The state in those days also performed a regulatory function. Entrepreneurs were forbidden to deal in the transport of slaves, or to send young children up chimneys and down coal mines. Factory Acts introduced legislation about health and safety. It was taken for granted that humanitarian concerns could never be safely entrusted to self-regulation by the private sector.

Neither could environmental ones. The founders of the new mines and factories were well aware that "where there's muck there's brass," but they themselves moved out of the muck as soon as practicable and directed operations from a distance. They could see no prospect of making money out of providing sewers or clean drinking water, so public health and safety were left to the public authorities. There grew up a division of labour between the state and private enterprise, a mutually advantageous system with the balance of power in the hands of the state.

Those days are over. Capitalism has now gone global. A multinational corporation no longer thinks of itself as subordinate to any government. It may build up an empire with greater financial resources than those of many independent nations, and multinationals now account for around a third of world output and two thirds of world trade. If their operations are impeded in the country where they originated, they can threaten to transfer their operations to a new site where wages are lower, taxes more avoidable, officials more bribable, and attempts to curb pollution non-existent.

That threat is not the only lever they control. In some democratic countries the financial cost of fighting an election have gone through the roof. Large contributions to party funds are frequently welcome even with strings attached. Governments base their claim to power on the fact that they represent public opinion, but any tycoon who can afford to buy up a portfolio of television stations and newspapers may well feel that he has public opinion in his pocket. He can certainly do serious damage to the election campaign of any party that antagonises him.

These global operators are determining interest rates, exchange rates, and the allocation of capital, irrespective of the political objectives of national political leaders. Neo-conservatives approve of this. They urge that governments should be content to function as the agents of these organisations—that is, for as long as governments continue to exist. The more confident among them believe that that will not be for very long. Here is a selection of sentiments from some of their thinkers:

"Like a mothball, which goes from solid to gas directly, I expect the nation-state to evaporate...There will be no more room for nationalism than there is for smallpox."

"In a borderless economy, the nation-focussed maps we typically use to make sense of our economic activity are woefully misleading...the old cartography no longer works."

"The nation state is dead."

Its death is heralded as the confirmation that the End of History has now been reached. Should we rejoice at this? After all, the Left in the past has directed many of its fiercest criticisms against the power and parochialism of nation states. It dreamed of a world where the boundaries between them might melt away. So why are radicals like Noam Chomsky apparently changing their minds and writing things like: "My short term goals are to defend and even strengthen elements of state authority"? It is because "Unlike the private tyrannies, the institutions of state power and authority offer to the despised public an opportunity to play some role, however limited, in managing their own affairs." The end of history as envisaged by the new prophets would involve the end of democracy. Of course, nobody imagines that a manifesto saying "Down with democracy" would be greeted with cheers. But you can sidestep that little difficulty by saying instead: "Down with the only institution that makes democracy feasible."

Many things can be claimed for the market as an indispensable way of regulating the supply and demand of goods and services. It has transformed our world and we cannot do without it. But nobody claims that it is democratic. It has never pretended to believe that all men are created equal or should be treated as if they were equal. It believes only that all dollars are created equal. It follows that more time and effort should be devoted to satisfying the whims of a rich man than to saving a poor one from starvation. In the past it has functioned well as a partner of the state, but stripped of all restrictions it could make a disastrously bad master.

Experience already shows that in those countries which have most eagerly accepted the neo-conservative gospel, the gap between the richest and the poorest has widened and continues to widen. In the United States it has brought about a level of inequality unknown since the 1920s. By 1990 Chief Executive Officers in America earned roughly 150 times the average worker's salary, while in Japan the figure was 16 and in Germany 21. In Europe the greatest degree of inequality is in England, where the same economic gurus that preached Reaganism were revered by Thatcher and later by Blair as the pinnacles of economic wisdom.

The social effects of this growing gap are far-reaching. Laissez-faire capitalism offers rich rewards for the winners, but only governments can provide a safety-net for the losers. They must try to keep taxes low in order to win elections and avoid frightening away big business. Yet it gets harder to keep taxes low for the law-abiding wage-earner, because the rich and powerful have an aversion to paying any taxes at all. "Only the little people" do that. Information about individual tax payments is normally unavailable, but in 2000 a newspaper obtained a copy of the tax returns of one British-based multi-millionaire and revealed that in the previous year he had paid only £5000 in tax.

The softest target for tax cutting is welfare. In the economists' vocabulary welfare has become a dirty word, with the implication that most of the people who might want to claim it are no-hopers anyway, and not really trying to compete. The danger is that some of them may get the message that in this society only money can command respect, and may therefore decide to go out and steal some, or demonstrate their entrepreneurial spirit by a spot of drug dealing.

The response to that is to keep them in jail. In California, the jail population became eight times greater within a single generation. By the start of 1997 around one in fifty American males was behind bars and one in twenty on parole or probation. Blacks are seven times likelier than whites to be imprisoned and one in seven black men has been imprisoned at some point in his life. In 2004, according to a report in the journal *American Prospect*, "people die younger in Harlem than in Bangladesh."

These rates of incarceration are ten times the rates in most of Europe. But they are still not high enough to convince the population at large that they are safe in their own houses, and something like thirty million Americans—over ten per cent of the population—have withdrawn into gated, privately guarded buildings or housing developments. Britain, converted to the same economic principles, by now has a rate of imprisonment way below America's but far higher than that of any other European country and is rising fast.

Gated communities, a relatively new phenomenon in this country, are now increasingly featuring on the estate agents' prospectuses. The crystal gazers commend these trends on the grounds that the increasingly wide difference in life-styles represents an ever greater range of "choice."

They have constructed a road map of the future depicting one road which goes straight on to the horizon with no turnings, and all the signs on it read: "No Stopping. No turning back." This vision is popular with politicians. It is easy to understand and lends itself to exhilarating flights of oratory like Thatcher's "The lady's not for turning" and Tony Blair's "I have no reverse gear."

Taken at its face value, it suggests that from now until the end of time things can only move in one direction. The index of economic growth will continue to go up and up and that will make people happier and happier. More and more of the world's natural resources will be converted into consumable goods, which must be endowed with built-in obsolescence, since it is vital that they are shunted at ever-increasing speed from assembly belt to landfill site or recycling depot in order to make room for more. More people will be in jail. More people will spend more of their time moving from one place to another and putting ever-higher premiums on the speed at which they travel. Less and less of the price of an article will be concerned with the costs of production, and more and more will be spent on persuading people that they want it. Obesity is only one outward and visible sign of the process of force-feeding them with wants, and conditioning them to feel ashamed if they fail to satisfy them.

The forecasts accept that the distribution of wealth will become ever more polarised, but Professor Hayek taught us not to worry about this apparent injustice. It would indeed "have to be regarded as very unjust *if* it were the result of a deliberate allocation to particular people." But it should be explained to the losers that there is nothing personal about the process, and therefore resentment would be quite out of place. The workings of an economic law "cannot be just or unjust." It merely feels that way sometimes.

For some years after the phrase "The End of History" was coined, the prophecies that flowed from it were apparently being fulfilled. They had been arrived at quasi-scientifically, by extrapolating some of the current graphs showing the ongoing expansion of economic growth. They skimmed over the fact that two other graphs had been climbing over the same period of time. One was the graph for the consumption of fossil fuels on which the productivity depended. The other was the graph of world population on which it relied for a bottomless pool of cheap labour and an ever-growing army of consumers. Both of these are due for a nose-dive in the near future.

There will always be reserves of fossil fuel but the time is coming when the energy they yield will be less than the energy it requires to access them, and no amount of investment can repeal the laws of thermodynamics. Renewable sources will be found, but relying on them will be more like living on income and less like indulging in *la dolce vita*, as we have recently been doing, on the proceeds of Nature's long-hoarded capital.

Secondly, by counting the number of babies already born, it is unanimously concluded that within decades the total world population will cease to rise. This statement is often followed by

"...and will then level off." It is hard to see what the levelling off prophecy is based on. The strong social, technical and biological factors conducive to the fall in the birth rate are likely to remain in force. We cannot un-invent birth control, nor wipe the memory of it from our collective consciousness. Market forces if unchecked will continue to suck women into the labour market as they do now.

It seems equally on the cards that the peak will not be followed by a levelling off, but will prove to be the high point of a parabola, and that the numbers will proceed to fall as steadily as they previously rose. That could be a good thing in the long run for our own species as well as many others, but the transitional period will call for some radical changes in our sense of values.

Such ideas were once denounced as "declinism" and blamed on left-wing propaganda—until 1998, when John Gray, Professor of European Thought at the London School of Economics, published a book entitled *False Dawn: The Delusions of Global Capitalism*. Robert Frank in America had already published a powerful critique of its emergent "Winner-Take-All" society, but Gray was more concerned with attacking the prophecies enshrined in the neo-con prospectus.

He concluded that sooner or later "the jerry-built edifice of global laissez-faire will begin to crumble" and in the interim "social exclusion and political alienation will be a constant danger." The Left was quite ready to believe that, but Gray was offering nothing for their comfort. They were also instructed "the collapse of socialism looks irreversible. For the future we can see, there will not be two economic systems in the world, but only varieties of capitalism." And it was already too late, according to

Gray, even to dream of a return to the soft left ideals of social democracy, as advocated by Keynes and Beveridge and Galbraith.

The Right was even more outraged by the assertion that the dream of a global free market—the avowed goal of the International Monetary Fund and other transnational agencies—is "not an iron law of historical development, but a political project." The effort to attain it "engenders new varieties of nationalism and fundamentalism even as it creates new elites." "Democracy and the free market are competitors rather than partners."

Despite the hostility it aroused, the book quickly became an international best-seller, translated into twelve languages. Three years later, the shock of 9/11 shattered the complacency of the neo-cons at the same time that it stiffened their determination. Gray's predictions of political alienation and the rise of extremist movements, which had seemed so wildly apocalyptic, could no longer be laughed off. But there was a touch of Cassandra about his forecasts. He offered no way out. At one point he referred disparagingly to "the conviction that humankind's ills can be cured by an act of will," and you can't get much gloomier than that. It's not even true. Not all of our ills can be cured by acts of will, but an awful lot of them have been.

However, in his final chapter he did strike a slightly more positive note. "A basic shift in economic philosophy is needed," he wrote. "The freedoms of the market are not ends in themselves. They are expedients, devices contrived by human beings for human purposes. Markets are made to serve man, not man the market." He leaves a question mark over the possibility that our inherited "resources of critical rationality" might enable us to think our way out of our problems.

But if we seriously accept that the market was made for man, the question marks will grow bigger than that. We can decide that if it is not working to our satisfaction, we can adjust it. It is not enough to declare as an eternal verity that we can never return to the principles of social democracy practised in the last century, which some prosperous European countries have never abandoned. It is like saying: "You have made your bed and now you must lie on it"—one of the stupider adages. What's to prevent you from getting up and making it again, having learned from your mistakes? In the past when the market malfunctioned, a return to those principles swiftly followed, as they did in America after 1929.

Depressingly, it is not only the neo-cons that are sold on the new fatalism. It has seeped into left-wing thinking as well. Here is a passage from an article in the *New Statesman:*

"The broad signs are that higher degrees of wealth concentration are proving more acceptable than in the past. If so, Britain is almost certainly in the early stages of an epoch that accepts extreme personal fortune and growing inequality. The wealth explosion of the past two decades looks increasingly like a permanent shift, and one that is not yet complete. If so, it is the age of egalitarianism, which lasted for a generation and a half, that may come to be regarded as an aberration, an interruption of a more natural state of deeper economic and social polarisation."

"A *permanent* shift"—that's Fukuyama. "A more *natural* state"—that's Pinker. If even on the Left we really start believing in all this claptrap, there is nothing to be done but fasten our seat-belts and cover our eyes and wait for the road-map to run out of road.

This view of our predicament leads to the kind of paralysis of will among the well-intentioned that Yeats described:

"The best lack all conviction, while the worst
Are full of passionate intensity."

There is a lot of dejection on the Left. There is a fear that nothing can withstand the power now wielded by America; that the power of unfettered global capitalism is impregnable and its forward march unstoppable; that humanity's hopes will always founder on the allegedly "deeper bedrock" of our greed and aggression; that the reforms of the seventies are being stamped out by a new generation wallowing in self-indulgent consumerism; that the few still clinging to the old aspirations are scattered and isolated while the forces ranged against them consolidate into a world-wide hegemony. That is a deeply distorted picture. Let's try toning it down a bit.

<u>America.</u>

The USA is going through a phase of believing that the world is its oyster. Most of Europe consists of nations which have passed through that phase, from the rise of the Roman Empire to the eclipse of the British one, with a dozen others in between. It's worth hanging on to the reflection that anything the Americans are doing, we have done—and worse, and not so long ago either. Within living memory, the universal hate slogan "British go home" was as ubiquitous as "Americans go home" is now, and it seemed quite as futile. But the point is that like all our predecessors we did go home. And the Americans will too.

The Young.

Every generation revolts against the ideas of those who came earlier. Students can readily be taught to believe that their predecessors of a previous generation were a deplorable bunch, hung up on boring issues like parity of esteem. But the gains that were made are not under threat. They cannot for example be persuaded to polish up the concept of racism and wear it once again as a badge of honour.

Globalisation.

Multinational corporations are not a force of nature. They are made of people. They may bribe researchers to report that nicotine—(or sugar, or jumbo-burgers)—turns out to be really good for you after all. But whistle-blowers give the game away, and governments, when the chips are down, have to be seen to be on the side of the customers.

Corporations, however mighty, depend absolutely on our colluding with them by conforming to the "gotta-have-it" role they prescribe for us. The three words that frighten them most are "No, thank you." It worked on a small scale when Martin Luther King brought a bus company to its knees by asking his followers to say "Not at any price" to its segregated service. On a larger scale, Nescafe was baffled when its efforts to screw repayments of old debts out of famine-stricken Ethiopia led to a drop in sales: what business was that of coffee-drinkers?

Indeed, it's an empowering phrase that's been known to cause empires to shudder and fall—as when Gandhi said "We don't have to buy your salt," or Bostonians said "We don't have to buy your tea." Some day, if sufficiently motivated, people might even revolt against the latest gotta-have-it mantra: "Live now, pay later. " They might start saying: "No thank you, we'll wait. We don't have to borrow your money."

The Road Map.

Meanwhile the confidence in the onward-and-upward advance of Winner-Take-All Economics is wavering. The blatant luxury fever reminds some reputable commentators of Versailles and *apres moi le deluge*. Others recall the South Sea Bubble, and the similar fever that preceded the Wall Street crash. High street sales are being underpinned by measures which can only be temporary. In Britain a crazy inflation of house prices and aggressive selling of second mortgages encourages home-owners to spend more money in a year than they will earn in two or three. America, the richest country in the world, sinks deeper and deeper into hock to its poorer neighbours. Fears have been expressed that the vast expansion of private indebtedness may "proletarianise" the middle classes by destroying the sense of security they once enjoyed. The powers that be may not be installing reverse gears, but they are very nervously checking their brakes.

Human Nature.

The first essential is to throw out the myth that compassion is less deeply rooted in our evolved nature than anger and hatred, and "you can't change human nature." People have the capacity to be vile, and the capacity to be co-operative; most of the time most of them are not vile, and if we take a long enough view they seem to be getting better rather than worse.

Steven Pinker, by no means a fan of the species, cannot help noticing that nowadays most people live their adult lives without pressing their violence buttons. He points to a hundred-fold decrease in homicide since the Middle Ages, a general decline over the last century in the taste for capital and corporal punishment, and notes that Western sensibilities of late have "steadily recoiled from the glorification of combat."

We could seek specific reasons for the more recent changes, but the process of abandoning behaviour patterns because they are stupid or disgusting has been going on for a very long time. One of the earliest presumably was cannibalism. There is growing evidence that originally it was what *Homo sapiens* did. Human flesh is tasty and nutritious—dead children are particularly succulent—so if there was a corpse around, why would anybody want to waste good food? Yet in most of us today, the thought of eating people causes as powerful a reaction of physical disgust as the thought of eating excrement. How did that come about?

It was probably the women who started it. They were the most likely to have an excess of the empathic hormones: rage at the thought of anyone eating their children could spread to embrace other people's. The ones who felt most strongly about it would be the ones best at fostering, and so would leave more surviving children, and most young animals pick up their ideas of what is edible or not edible from their mothers, so both genes and culture would be pulling in the same direction. In ways like this, human behaviour can and does change, by exhortation and example. Somewhere in the world the last cannibal ate the last meal of long pig. Somewhere in Europe, without knowing it, two gentlemen in a field at dawn fought the last duel. Somewhere in the Southern states, without being aware of it, the last lynch mob hanged its last victim.

In isolated areas cannibalism survived to our own day. So, in some rural parts of South East Europe, did the practice of vendetta. Under that system, if you killed a member of my family it was my duty to kill one of yours—then it was your turn again, on down the generations. Groups of do-gooders with no axe to grind went there and pointed out that an eye for an eye leaves the whole world blind and that in most parts of the world people have abandoned this habit. They made their point. It's not too difficult for the human mind to grasp. It may well be that the last honour killing has already taken place.

To an unbiased onlooker it would seem that the same vicious circle is operative in Palestine. Every act of violence is justified by its perpetrators as retaliation, and greeted by its victims with the vow that "This cannot be allowed to go unpunished." But in that arena, appeals to reason are futile. One problem is the overriding influence of the supernatural. The addiction to vendetta was hard to break, but at least the assassins did not claim that they were carrying out the will of God. In the Middle East, however, the fundamentalists of all three monotheistic religions seem to have moved into the driving seat, even though the fundamentalists in all three are in a minority. As a child I was familiarised with all the words attributed to Jesus of Nazareth, the prince of peace. I find it hard to imagine the Christ of the New Testament contemplating the scores of thousands of Iraqi corpses and listening to the zealots who explain: "There's been some collateral damage here. But give us your blessing, Lord: it is all being done in your name."

War is much harder to glorify than it used to be, but one effective way of glorifying it is to say that you are fighting for good against evil. It is more effective still if you make yourself believe it. George Bush is fond of using those terms. When using them he implies that an evil ruler is one who stockpiles weapons of mass destruction and refuses to let UN inspectors examine them; who invades other countries and drops bombs on them; who damages world trade by providing subsidies and imposing embargoes; who pollutes the planet and sanctions any means deemed likely to secure victory—always with the proviso that none of these things is evil if it is done by America. He is not unusual in this. All combatants assert that their cause is just.

It is tempting to think that if religion became obsolete, the problems would be easier to solve—but that is simplistic. Ideologies can lead to an equally manic frame of mind. In terms of science fiction, if some wise aliens were to alight on our planet they would conclude that *Homo sapiens,* due to some flaw in the blueprint, has evolved very little resistance to this lethal fever of mutual slaughter, which twice in the last century built up into a world-wide pandemic. The prospect that some day the last war will have been fought looks exceedingly remote.

We cannot expect the arrival of the visitors from outer space to point out the futility of this way of conducting relationships. But a new voice may be beginning to emerge as our globe becomes a smaller place: the voice of world opinion. Most of the earth's inhabitants are able to make a dispassionate assessment of events in which their immediate interests are not involved, and most of them can take a relatively detached view of events in the Middle East. The rival claims to be fighting for Good against Evil may play well on the home front, but in the sight of those who believe in quite different Gods or none at all, such rhetoric appears dangerous and deranged. Unlike during the Cold War, governments do not show much appetite for picking a side and lining up behind it—and the people they govern are even more reluctant. The natural response is "a plague on both your houses," a longing for a *cordon sanitaire* to prevent the malady from spreading, and criticism of anyone who appears to be upping the ante.

It has recently become much harder, especially in the democracies, to remain totally deaf to world opinion. The United Nations is as fallible as any other human institution. It can be painfully slow in arriving at an agreed conclusion, and may then be powerless to enforce it. But it's there, it's evolving, it cannot be ignored, and even in its present form it makes the world a perceptibly safer place.

One event in recent years was more dramatic than any debate in the chambers of the U.N. On the eve of the invasion of Iraq, many millions of people in countries all over the world, of different races, generations, creeds, and cultures, went out into their streets and public places, and stood in the cold or the heat or the rain to express non-violently their conviction that the projected move was a step in the wrong direction. It was spontaneous. It was unprecedented. That conviction has not diminished, and not all the criticism has come from neutral countries. In America and in Great Britain, public opinion has been split right down the middle over the issue. In the American election of 2004 Bush carried the day, but the queues at the polling stations were as long as they once were in South Africa, and cast a new light on the theory that apathy and acceptance are here to stay.

Can we look forward to the time when the last war will have been fought? Probably not in our time. But a change of perception is under way. Professor Pinker surmised that in due course human ethical attitudes might be expected to move another step up the ethical escalator, but he could see no clear sign of what form it might take.

Perhaps he was looking in the wrong direction.

◆ ◆ ◆

REFERENCES

Dedication

"But principally I hate and detest..." Jonathan Swift. 1725. In a letter to Pope, 29th September.

Chapter 1—*The Invitation*

"The greatest enterprise..." E. O. Wilson. 1998. *Consilience: The Unity of Knowledge*. Little, Brown. p. 6.

"When I find myself in the company..." W. H. Auden. 1963. *The Dyer's Hand*, "Poet and the City."

"...how I and many other scientists feel..." Richard Dawkins. 1998. *Unweaving the Rainbow*. Penguin Press. p. 15.

"...an invitation." 1998. C*onsilience: The Unity of Knowledge*. Op. cit.

Chapter 2—*Darwin and God*

"Man has never been the same..." E. St.V. Millay. *Conversation at Midnight*, IV.

"My working men..." T. H. Huxley. Quoted in A. Desmond. 1994. *Huxley*. Penguin. p. 292.

"Man has worked his way..." T. H. Huxley. 1893. Essay on *Evolution and Ethics*.

Chapter 3—*Darwin and Marx*

"Darwin recognises among beasts and plants..." Quoted in Adrian Desmond and James Moore. 1991. *Darwin*. Michael Joseph, Ltd. p. 485.

"The point is, to change it." Karl Marx. 1888. *Theses on Feuerbach*.

"Darwin's book is very important..." Francis Wheen. 1999.

Karl Marx. Fourth Estate. p. 364.

"I thank you for the honour..." Ibid. p. 363.

"If the misery of the poor..." Quoted in Steven Pinker. 2002. *The Blank Slate: The Modern Denial of Human Nature.* Allen Lane. (Penguin Books.) p. 151.

"I would prefer the Part or Volume not to be dedicated..." 1999. *Karl Marx.* Op. cit., p. 365.

"The two atheists were invited to lunch..." Desmond and Moore. 1991. *Darwin.* pp 656-7.

Chapter 4—*Lamarckists*

"...abominable trash vomited forth..." P. Corsi. 1988. *The Age of Lamarck: Evolutionary Ideas in France, 1790-1834.* Berkeley: University of California Press. chs 3-4.

"...most beautiful and wonderful..." From the closing sentence of Darwin's *The Origin of Species.*

"What a book a Devil's Chaplain might write..." Charles Darwin (1856) in a letter to J. D. Hooker.

"When its whole significance dawns..." G. B. Shaw. 1965. *The Complete Prefaces of Bernard Shaw.* Paul Hamlyn. p. 520.

Butler on Darwin...Ibid., p. 523.

"...literary people have for some reason felt themselves entitled to express..." Peter Medawar. 1996. *The Strange Case of the Spotted Mice.* O.U.P. p. 198.

"Weismann proposed as an immutable law..."August Weismann. 1893. *The Germ-Plasm: A Theory of Heredity.*

"Richard Dawkins has described..." 1987. *The Blind Watchmaker.* W. W. Norton.

"...nevertheless inherited by subsequent generations." Michael Balter. 2000. Was Lamarck just a little bit right? *Science,* 288:38.

Chapter 5—*The Twig Is Bent*

" 'Tis education forms..." Alexander Pope. 1735. Ep ii. *To a Lady*.

Oliver Twist. R. C. Lewontin. 1991. *The Doctrine of DNA: Biology as Ideology*. Penguin. p. 23.

"...denounce contemporary attitudes..." 2003. *The Blank Slate*. Op. cit., p. 6.

Henry Harlow. Affectional Responses in the Infant Monkey. *Science*, 130: 421.

John Bowlby. 1991. *Attachment and Loss, Vol 2: Anxiety and Anger*. Penguin. p. 162.

Chapter 6—*Thicker than Water*

"At times I was almost sure I saw..." W. D. Hamilton. 1996. *Narrow Roads of Geneland*. W. H. Freeman. p. 25.

"Wynne Edwards was convinced..." Wynne Edwards. 1962. *The Evolution of Social Animal Dispersal in Relation to Social Behaviour*. Oliver and Boyd.

"George Williams relieved his feelings by writing a book." G. Williams. 1966. *Adaptation and Natural Selection*. Princeton University Press.

"The rejection of group selection was celebrated..." E. Sober and D. S. Wilson. 1998. *Unto Others*. Harvard University Press.

"Sir Solly Zuckerman admitted..." 1991. *The Doctrine of DNA*. Op. cit., p. 9.

"Hamilton...succeeded in getting it published." W. D. Hamilton. 1964. The genetical theory of social behaviour 1 and 2. *Journal of Theoretical Biology*, 7, 1-16; 17-32.

"E. O. Wilson gave a vivid account..." 1994. *Naturalist*. Island Press. pp 319-20.

"He believed that all social insects..." 1996. *Narrow Roads of Geneland*. Op. cit., pp 388-9.

"...must endure the tortures of Orestes..." Ibid., p. 189.

Chapter 7—*Bread upon the Waters*

"...the whole field of animal behaviour was stampeding..." Ullica Segerstråle. 2000. *Defenders of the Truth.* O.U.P. pp 84-95.

"...this may really be hot!" Ibid., p. 80.

"He called this process Reciprocal Altruism." R. L. Trivers. 1971. The Evolution of Reciprocal Altruism. *Quarterly Review of Biology*, **46**, pp 35-7.

"...the success rate would start to decline..." Richard Dawkins. 1976. *The Selfish Gene.* O.U.P. p. 220.

"...dragged kicking and screaming..." Matt Ridley. 1996. *The Origins of Virtue.* Viking. p. 56.

"...the baby's eye view..." R. L. Trivers. 1974. Parent-offspring conflict. *American Zoologist*, **14**, pp 249-64.

"...honoraria..." 1996. *Narrow Roads of Geneland.* Op. cit., p. 185.

Birds and ticks. 1976. *The Selfish Gene.* Op. cit., p. 183.

Vampire bats. G. S. Wilkinson. 1984. Reciprocal food-sharing in the vampire bat. *Nature*, **308**, pp 181-4.

Chapter 8—*The Troubles*

"There is politics aplenty..." Alper et al. 1976. The implications of Sociobiology. *Science*, **192**, 424-5.

"People who brandish naturalistic principles at us..." P. Medawar. 1960. *The Future of Man.*

"...no interest in the background of the author..." Richard Feynman. 1998. *The Meaning of It All.* Allen Lane. p. 22.

"Our rhetoric was at fault." R. Lewin. The course of a controversy. *New Scientist*, 13th May, 1976. pp 344-5.

"I was raised as a racist..." E. O. Wilson, interview in *The*

Guardian, 17th Feb, 2001.

"Unusually enough, this dispute has now been resolved." A. Brown. 1999. *The Darwin Wars*. Simon and Schuster UK, Ltd.. p. 64.

"It may be a hasty conclusion..." 2000. *Defenders of the Truth*. Op. cit., p. 311.

Chapter 9—*Genes and Memes*

"Let us try to teach..." 1976. *The Selfish Gene*. Op. cit., p. 3.

"I am almost driven to the despair..." 1998. *Unweaving the Rainbow*. Op. cit., p. x.

"Such a very proper purging..." Ibid., p. *ix*.

"A hen is merely..." Samuel Butler. 1912. *Note Books*. Ed. H. Festing Jones. ch. 1.

"...should be thought of as a vehicle..." Richard Dawkins. 1982. *The Extended Phenotype*. O.U.P. p. 8.

"The Necker Cube model is misleading." Richard Dawkins. 1989 edition. Preface to *The Selfish Gene*. p. *ix*.

"Memes should be regarded as living structures." Ibid., p. 192.

"...accused of having backtracked.." R. Dawkins. 1998. *The Devil's Chaplain*. Allen Lane. p. 126.

"We do not say 'It's as if I have intentions'..." Susan Blackmore. 1999. *The Meme Machine*. O.U.P. p. 229.

"The books, the telephones, the fax machines..." Ibid., p. 204.

"I am not initially attracted..." Daniel C. Dennett. 1995. *Darwin's Dangerous Idea*. Simon and Shuster. p. 346.

"...brains seem to be designed to transform, invent.." Ibid., p. 355.

Chapter 10—*The Pleistocene Inheritance*

"The mind therefore consists of...." Matt Ridley. 2003. *Nature via Nurture*. Fourth Estate. p. 63.

The seminal ev/psych volume. *The Adapted Mind*. 1992. J. H. Barkow, L. Cosmides, and J. Tooby (eds.) Oxford University Press.

"...to gouge out an eye..." Raymond Dart, to camera. Reproduced in the film *The Aquatic Ape*, produced by BBC Natural History Unit for the television channel *Discovery*.

"Wilson felt there was nothing new..." 2000. *Defenders of the Truth*. Op. cit., p. 317.

"Its capacity to learn to speak..." N. Chomsky. 1975. *Reflections on Language*. (Pantheon Books) Random House.

"EEA is not a particular place..." From the EP Website.

Philip Kitchen. 1985. *Vaulting Ambition*. MIT Press.

"The aesthetic preference for symmetry..." John Alcock. 2001. *The Triumph of Sociobiology*. O.U.P. p. 138.

"...a beauty-detection mechanism designed specifically..." Randy Thornhill and Craig T. Palmer. 2000. *A Natural History of Rape*. MIT Press. p. 71.

Chapter 11—*Cinderella*

Martin Daly and Margo Wilson. 1998. *The Truth about Cinderella: A Darwinian View of Parental Love*. Weidenfeld and Nicolson. Passim.

Chapter 12—*Rape*

"Mistaken notions about what causes..." 2000. *A Natural History of Rape*. Op. cit., p. xi.

"We will refer to it as the 'social science' explanation." Ibid., p. 123.

"...the environmental problems our ancestors faced..." Ibid., p. 17.

"...the widespread occurrence of rape across animal species..." Ibid., p. 146.

"Craig Palmer found rape to be..." C. Palmer. 1989. Rape in non-human species: definitions, evidence, and implications. *Journal of Sex Research*, **28**: 353-374.

Chapter 13—*The Origin of Empathy*

"Although male and female researchers..." Sarah Blaffer Hrdy. 1999. *Mother Nature: Natural Selection and the Female of the Species.* Chatto and Windus. p. 53.

"The important point here..." Ibid., p. 56.

"Richard Dawkins sees no reason..." 1982. *The Extended Phenotype.* Op. cit., p. 57.

"Loving arose in many cases..." Konrad Lorenz. 1966. *On Aggression.* Methuen. p. 186.

Robert Frank. 1998. *Passions within Reason.* Norton.

"With the advent of parental care..." 1971. I. Eibl-Eibesfeldt. *Love and Hate.* Methuen. p. 123.

"Frans de Waal gives examples..." Frans de Waal. 1996. *Good Natured: The Origins of Right and Wrong in Humans and Other Animals.* Harvard U. P. Passim.

"One female spider monkey..." Alejandro Estrado. 1982. A case of adoption of a howler monkey infant by a female spider monkey. *Primates,* **23**(1)*:* 135-137.

"The endocrine equivalent of candle-light..." 1999. *Mother Nature.* Op. cit., p. 154.

"More important to animals which form long-term bonds..." C. Carter, Sue and Lowell L. Getz. 1993. Monogamy and the prairie vole. *Scientific American,* **268**. pp 100-106.

"Female humans are like that already..." 1996. *The Origins of Virtue.* Op. cit., p. 169.

"There is no appreciable quid pro quo..." Erving Goffman. 1997. The arrangement between the sexes. *Theory and Society.* **4** (3):301-332

Chapter 14—*It's a Boy*

"Better fighters..." Richard Wrangham and Dale Petersen.

1997. *Demonic Males*. Bloomsbury.

"Dart, who outlined..." Raymond Dart. 1953. The Predatory Transition from Ape to Man. *International Anthropological and Linguistic Review*, Vol 1. No. 4.

Marlene Zuk. 1993. Feminism and the study of animal behaviour. *Bio Science*, **43**(11): 774-778.

Pascal Gagneux, David S. Woodruff, and Christophe Boesch. 1997. Furtive mating in female chimpanzees: comparable genetic studies from Gombe and from Sugiyama's study site at Bossou corroborate the Tai results. *Nature*, 387:327-8.

"Steve Jones confessed..." Steve Jones, 2002. *Y: The Descent of Men*. Little, Brown. p. 9.

"Hamilton's confession..." 1996. *The Narrow Roads of Geneland*. Op. cit., pp 190-191.

Chapter 15—*Right and Left*

"I think these arguments..." Jerry Fodor. 2000. *The Mind Doesn't Work That Way: The scope and Limits of Computational Psychology*. MIT Press.

"No one can make sense..." 2003. *The Blank Slate*. Op. cit., p. 284.

"With divisive moral issues..." Ibid., p. 281.

"...really do vindicate..." Ibid., p. 293.

"I would say it is just sensible..." Interview published in *The Observer*, 22nd September 2002.

"...our coldly logical emotions..." Steven Pinker. 1997. *How the Mind Works*, Allen Lane. pp 404-5.

"The Banker's Paradox." Ibid., p. 507.

"...the declining band of Skinnerians..." W. G. Runciman. 1998. *The Social Animal*. Harper Collins. p. 55.

"...constant interaction..." 1991. *The Doctrine of DNA*. Op. cit., p. 26.

Philip Roth. 2000. *The Human Stain*. Jonathan Cape.

"When we began..." Richard J. Herrnstein and Charles Murray. 1996. *The Bell Curve: Intelligence and Class Structure in American Life*. Free Press paperbacks. p. 554.

"...did indeed vividly..." 2000. *Defenders of the Truth*. Op. cit., p. 309.

"...drew a graph. 1996. *The Bell Curve*. Op. cit., p. 222.

"Its name is trade." 1996. *The Origins of Virtue*. Op. cit., p. 193.

"...the lack of clear property rights..." Ibid., p. 239.

"Benevolence...should be subject to the greatest scrutiny." Speech by Roger Kimball reported in *The Guardian*, 31st May, 2003.

"...a best-selling book..." Ann Coulter. 2003. *"Treason: Liberal Treachery from the Cold War to the War on Terrorism."*

Chapter 16—*Striding the Blast*

"To appreciate what has happened..." 2003. *Nature via Nurture*. Op. cit., p. 4.

"Nature versus Nurture is dead." Ibid., p. 280.

Frans deWaal. 1996. *Good Natured*. Op. cit., pp 178-80.

Chapter 17—*Progress*

"But as brains became more highly developed..." 1989. *The Selfish Gene*. Op. cit., p. 60.

"...not just incidentally progressive..." 2003. *A Devil's Chaplain*. Op. cit., p. 211.

"...(apes) have fewer oxytocin receptors..." 2003. *Nature via Nurture*. Op. cit., p. 48.

Chapter 18: *What's Left?*

"So why is the left...?" from the Evolutionary Pyschology website.

"A booklet by..." Peter Singer. 1999. *A Darwinian Left: Politics, Evolution, and Co-operation*. Weidenfeld and Nicolson.

"It often unmasks the universal hypocrisies..." Irwin Silverman. 2003. *Evolutionary Psychology*, March, 1: 1-9.

"The Flynn effect..." 1996. *The Bell Curve*. Op. cit., pp 307-9.

"Robin Dunbar was the first..." Robin Dunbar. 1996. *Grooming, Gossip, and the Evolution of Language*. Faber and Faber.

Erving Goffman. 1959. *The Presentation of Self in Everyday Life*. Doubleday.

"Contemporary hunter-gatherer societies..." D. Erdal and A. Whiten. 1996. Egalitarian and Machiavellian Intelligence in human evolution. *Modelling the Early Human Mind*. Eds. P. Mellars and K. Gibson. McDonald Institute Monographs, Cambridge.

"Research findings have..." Richard Wilkinson. 2000. *Mind the Gap: Hierarchies, Health, and Human Evolution*. Weidenfeld and Nicolson.

Chapter 19—*The Magic Hand*

"This book is about..." 2003. *The Blank Slate*. Op. cit., p. viii.

"Thomas Sowell's book..." T. Sowell. 1987. *A Conflict of Visions: Ideological Origins of Political Struggles*. Quill.

"He rebukes S. J. Gould..." 2003. *The Blank Slate*. Op. cit., p. 125.

"A community blood bank..." Robert Kuttner. 1996. *Everything For Sale*. University of Chicago Press. p. 65.

"Singer's book..." P. Singer. 1981. *The Expanding Circle: Ethics and Sociology*. Yale University Press.

"...a moral gadget." 2003. *The Blank Slate*. Op. cit., p. 270.

Judith Rich Harris. 1998. *The Nurture Assumption: Why Children Turn out the Way They Do*. Free Press.

Howard S. Schwartz. 2001. *The Revolt of the Primitive: An Inquiry into the Roots of Political Correctness*. Praege.

"The liberty of the strong must be..." Sir Isaiah Berlin. 1969. *Four Essays on Liberty*. O.U.P. p. 124.

"A book entitled..." John Micklethwaite and Adrian Wooldridge. 2003. *The Company: A Short History of a Revolutionary Idea*. Weidenfeld and Nicolson.

"...crony capitalism..." Examples are described in Amy Chwa. 2003. *World on Fire*. Heinemann.

Chapter 20—The End of History?

"...at present the received opinion..." William Morris quoted in Noam Chomsky. 1996. *Powers and Prospects*. Op cit., p. 74.

"The End of History." The phrase first appeared in an article by Fukuyama in *National Interest* in the summer of 1989. Fukuyama later used it as a book title: *The End of History and the Last Man*. The Free Press. 1995.

"I am not going to fall into the trap..." 1998. *The Origins of Virtue*. Op. cit., p.260.

"Like a mothball..." Nicholas Negroponte. 1995. *Being Digital*. Hodder and Stoughton.

"In a borderless economy..." Keniche Omae. 1995. *The End of the Nation-State*. Harper Collins. pp 19-20.

"The nation state is dead." John Naisbitt. 1995. *Global Paradox*. Nicholas Breasley Publishing. p.40.

"My short-term goals are to defend..." 1996. *Powers and Prospects* Op. cit., pp 73-74.

"...die younger in Harlem than in Bangladesh." Lawrence Jacobs and James Morone. 2004. *American Prospect.*

"Hayek taught us...".Friedrich von Hayek. 1994. *The Road to Serfdom.*

"Winner-Take-All..." R. H. Frank and Philip Cook. 1995. *The Winner Take All Society.* The Free Press.

"A basic shift is needed..." John Gray. 1998. *False Dawn: The Delusions of Global Capitalism.* p. 234.

"The broad signs are that..." Stewart Lansley. 4th October 2004. *The New Statesman.* p.29.

"The best lack all conviction..." W. B. Yeats, in his poem *The Second Coming.*

"...contemplating the scores of thousands of Iraqi corpses and..." Les Roberts, Riyadh Lafta, Richard Garfield, Jamal Khudhairi, Gilbert Burnham. 13th November 2004. Mortality before and after the 2003 invasion of Iraq: cluster sample survey. *The Lancet,* Vol. **364**, No. 9447.

♦ ♦ ♦

INDEX

A

Acheulean, 204
Adam Smith Institute, 251
The Adapted Mind: Evolutionary Psychology and the Evolution of Culture, 101
adaptive evolution, 210
adrenaline, 144
agnostic, 17
Alper, J., 75, 237
altruism, 49
America, 266
American Association for the Advancement of Science, 78
Archimedes, 226
Arrow, Kenneth, 250
athletic scholarships, 183
Attenborough, David, 88
Auden, W. H., 9
Aveling, Edward, 24–25
Axelrod, Robert, 64, 72

B

Bangladesh, 252
Banker's Paradox, 177
The Bell Curve, 180, 181, 184, 226
Belloc, 208
Berkeley, 85
Berlin, Sir Isaiah, 250
Betzig, L., 99
birth control, 263
Blackmore, Susan, 95
Blair, Tony, 259, 261
The Blank Slate, 10, 11, 237
blank slates, 42, 69
blood banks, 239
Boesch, Christophe, 167

brain
    as a computer, 10, 100
    as a gland, 100
Buchanan, James, 250
Burns, Rabbie, 204
Bush, George W., 178, 270
Butler, Samuel
    The Way of All Flesh, 33

C

canine teeth, 169
cannibalism, 269
capitalism, 257
Carter, Sue, 145
Cartesian, 177
Chief Executive Officers
    American, Japanese, German, 259
Chomsky, Noam, 78, 258
Cold War, 271
The Company, 250
conditioned reflex, 42
consilience, 9
Cosmides, Leda, 101, 108, 177
Creationists, 17, 224
Crick, Francis, 34
cystic fibrosis, 191

D

Daly, Martin, 111, 112, 115
Dart, Raymond, 103, 157, 239
Darwin, Charles, 22, 54, 173, 187, 207. See also *The Origin of Species*
    Christian society, 15
    death of, 17
    on progress, 30, 207

political impact, 17
Spencerian racism and, 24
theory of evolution, 14
Darwin, Emma, 21
Das Kapital
and Darwin, 23
Dawkins, Richard, 79–92
on poets, 9
The Extended Phenotype, 91
The Selfish Gene, 67, 209
the word "meme", 93
Unweaving the Rainbow, 88
de Vore, Irven, 59
de Waal, Frans, 198
Good Natured, 142
Demonic Males, 169
Dennett, Daniel, 72, 96, 219
Descartes, 178
A Devil's Chaplain, 210
Dickens, Charles, 39
DNA analysis, 166
Don Juans, 170
Donald Duck, 164
Donne, John, 218
dopamine, 211
downshift, 240
Dunbar, Robin, 228

E

economics, 259
inflation, 268
egalitarian ethic, 233
egalitarianism, 231
Eibl-Eibesfeldt, I.
Love and Hate, 138
empathy, 137, 140, 147
Enclosure Acts, 256
Enron, 249
environmentalists, 185
EP, 108, 129
EPC (extra-pair copulation), 166
Erasmus, 29
estrogen, 154, 155
estrus, 127

ethics, 77
ethology, 44
eugenics, 25, 26
Eureka, 226
eusocial, 214
evolutionary psychology, 101, 223
Expanding Circle, 241
The Extended Phenotype, 91

F

Factory Acts, 257
Faraday, Michael, 44
feminist, 124, 131, 132
Feynman, Richard, 81
Flynn Effect, 226
Fodor, Jerry, 173
fossil fuels, 262
four F's, 141, 148
Frank, Robert, 139, 263
Freud, Sigmund, 41, 137, 161
"subconscious", 40
Friedman, Milton, 250
Fukuyama, Francis, 265
The End of History, 255

G

Gagneux, Pascal, 167
Galileo, 225
Galton, Francis, 25
Game Theory, 71, 72, 137, 139
Gandhi, 267
gay gene, 156
Ghiselin, 137, 220
Gibbs, Lisle, 166
Gilbert and Sullivan, 174
glands, 211
globalisation, 267
Gods, 271
Goffman, Erving, 228, 229
Goodall, Jane, 229
gossip, 228

Gould, Stephen Jay, 56, 75, 77, 78, 81, 99, 164, 208, 238
Grant, Robert, 31
Gray, John
  False Dawn—The Delusions of Global Capitalism, 263
Great Depression, 249
group selection, 50

H

Haldane, J. B. S., 52
Hamilton, R. B. and R. M., 35
Hamilton, William, 49, 52, 53, 54, 59, 170
  inclusive fitness, 53
  kin selection, 55, 65
Hayek, F. A., 187, 242, 250, 262
Herrnstein, Richard J., 76, 180
Homo economicus, 239
homosexuality, 155
hormones, 144
Hrdy, Sarah Blaffer, 137, 145, 167
  Mother Nature, 144
Humphrey, N. K., 93
hunter-gatherers, 103
Huntingdon's chorea, 191
Huxley, T. H., 16, 54, 207
  on Darwinism, 31
  on evolution, 16
Hydraulic Model, 211

I

inclusive fitness, 53
infanticide, 114
instinct, 193
International Monetary Fund, 264
IQ
  Asians, 182
  Jewish, 182
  urban-rural dimension, 183

J

James, Oliver, 175
Jensen, Arthur, 76
Jones, Steven, 170, 210

K

Kamin, L., 78
Kansas City, 239
Keynes, J. M., 250
kibbutz, 215
Kimball, Roger, 187
kin selection, 55, 56
King, Martin Luther, 75, 243, 267
Kingsley, Charles, 15
Kitcher, Philip, 104
Koestler, Arthur
  Darkness at Noon, 33
Kuttner, Robert, 251

L

laissez-faire, 247
Lamarck, 29–31
Lamarckism, 51
Le Doux, 159
Lewontin, Richard, 39, 75, 77, 78, 89, 178
Limited Liability Acts, 256
London School of Economics, 52
Lorenz, Konrad, 137, 200
  ethology, 44
Lysenko, T. D., 34, 35

M

magnetic resonance imaging, 211
male bonding, 216
Malthus, Thomas, 187
  on human population, 22
Marat, Jean-Paul, 30
Marcos, Imelda, 252
Marx, Karl, 21, 23, 186, 224, 225

288 *Pinker's List*

and *The Origin of Species*, 22
Das Kapital, 24
daughter Eleanor, 24
mating behaviour, 104
McCarthy, Joseph, 187
McCarthyism, 179
Medawar, Peter, 33, 79
Mendel, Gregor, 54, 106, 173
Millay, E. St. V., 13
misogynism, 131
module, 106, 107, 129
monogamy, 167
moral escalator, 242
Morris, William, 255
multinationals, 257
Murray, C., 180

N

nation state, 258
National Health Service, 239
Nature via Nurture, 192
Neanderthals, 225
Necker Cube, 90, 91, 92, 97
neo-conservatives, 258
    9/11, 264
Nesse, Randolph, 87
neuro-transmitters, 211
New Statesman, 265
New Synthesis, 173, 188, 193
Not in Our Genes, 79
nuclear family, 117

O

olfactory resemblances, 215
orgasm, female, 127, 130
The Origin of Species, 15, 18, 54.
    See also Darwin, Charles
Owen, Wilfred, 201
oxytocin, 144, 145, 202

P

pain centre, 217
Palestine, 270
Palmer, C. T., 123–28, 132
Panglossian, 209
Pavlov, I. P., 40, 41
    conditioned reflex, 42
Peterson, Dale, 151, 169
phlogiston, 193
Pinker, Steven, 173, 185, 193, 237,
    238, 265, 268, 272
    "The Official Theory", 177
pity, 194
The Blank Slate, 10
pity, 194, 195, 201
pleasure areas, 212
Pleistocene, 102, 103, 104, 111,
    119, 127, 162, 256
Political Correctness, 178
Pope Pius XII
    on evolution, 17
posterity, 203
practitioners, 113
Price, George, 55
Prisoner's Dilemma, 63, 64, 71,
    139
progesterone, 154
progressivist bias, 208
protoplasm, 193
psyche, 211
psychology, 99

R

rape, 123, 124, 125, 132
    and beauty, 105
Reaganism, 259
receptors, 211
Reciprocal Altruism, 60, 61
relative deprivation, 233
Ridley, Matt, 64, 148, 191, 193,
    200, 255
    "the oxytocin story", 213

and trade, 184
Nature via Nurture, 192
The Origins of Virtue, 184
Roman Empire, 266
Roosevelt, Franklin, 249
Rose, Hilary and Steven
   Alas, Poor Darwin, 188
Rose, S., 78
Roth, Philip
   The Human Stain, 179
Rousseau, 178, 256
Rumsfeld, Donald, 178

S

San Francisco, 240
Schwartz, Howard S., 242
Segerstråle, Ullica, 59, 82, 181
The Selfish Gene, 85, 86, 89, 97, 184, 209
seratonin, 211
sexual dimorphism, 169
sexual harassment, 126
sexual selection, 151
Shakespeare, 194
Shaw, George Bernard, 29, 42
   support for Lamark, 32
shoaling, 197
Silicon Valley, 225
Singer, Peter, 241, 252
Skinner, B. F., 43
slave trade, 195
Smith, Adam, 187, 256
   The Wealth of Nations, 248
Smith, John Maynard, 35, 53, 56, 177, 199
   doubting Hamilton, 54
Snow, C. P., 9
Sobers, Elliott, 51
Social Darwinism, 26
social exclusion, 251
social science establishment, 124, 134
Sociobiology: The New Synthesis, 76

sociology, 77
Soviet Union, 224, 248
Sowell, Thomas
   A Conflict of Visions, 238
Stalin, Joseph, 34, 224
step-parenting, 111
Stevenson, Robert Louis, 88
Suharto, 252
supernatural, 270

T

testosterone, 144, 155
thalidomide, 153
Thatcher, Margaret, 49, 250, 259, 261
third fingers, 156
Thornhill, R., 123, 124, 128, 132
thumb-sucker, 159
Tinbergen, Niko, 44
Tit for Tat, 62, 64, 69, 71
tomboy syndrome, 157
Tooby, John, 101, 108, 177
trade unions, 244
Tragic Vision, 175, 184, 237
The Triumph of Sociobiology, 188
Trivers, Robert, 59, 61, 64, 76, 137
   preface to Dawkins's book, 79
Truman, Harry, 245
The Truth about Cinderella, 111, 113, 117
Turing, Alan, 173

U

United Nations, 272
Unweaving the Rainbow, 88
Utopia, 175

V

vampire bat, 65, 68, 69
vasopressin, 170
vulva, 160

W

Wall Street crash, 249
Wallace, Alfred Russell, 30
Walt Disney, 178
war, 245, 270, 271
Watson, J. B.
   blank slates, 42
Wegener, Alfred, 179
Weismann, August, 33
Wilberforce, William, 196
Wilde, Mrs., 157
Wilkinson, Richard
   Mind the Gap, 232
Williams, George, 50, 51
Wilson, D. S., 51
Wilson, E. O., 75, 76, 81, 99, 158, 192
   and evolutionary psychology, 103
   doubting Hamilton, 54
   on consilience, 9
   severe criticism of, 77
   Sociobiology, 77, 85
Wilson, Margo, 111, 112, 115
Women's Liberation, 75, 243
Woodruff, David, 167
Wordsworth, William, 88
world population, 262
Worldcom, 249
Wrangham, Richard, 151, 169
Wynne-Edwards, V. C., 50

Y

y chromosome, 151, 154, 170
Yeats, 266

Z

Zuckerman, Sir Solly, 53
Zuk, Marlene, 167, 168